DIVISION F

MATH PRACTICE WORKSHEETS

ARITHMETIC WORKBOOK WITH ANSWERS

More than 3100 division facts and exercises to help children enhance their elementary division skills

By Shobha

Table of Contents

Did You Know?

> Division is splitting into equal parts or groups. Symbol ÷ and / are used to denote division.

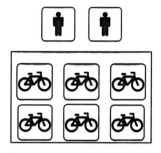 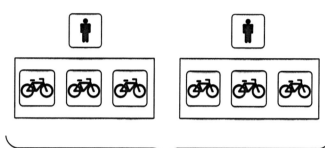

There are 6 bikes and 2 people. They both need an equal number of bikes.

They divided the bikes among themselves to get 3 each

In numbers: 6 ÷ 2 = 3 OR 6 / 3 = 2

> When you divide 0 by another number the answer is always 0. For example: 0 ÷ 2 = 0

> When you divide by 1 the answer is the same as the number you were dividing.
> For example, 2 ÷ 1 = 2

> When you divide a number by itself, you get 1.
> For example, 3 ÷ 3 = 1

> When you divide a number by 2, you are actually halving the number.
> For example, 6 ÷ 2 = 3

> When dividing, unlike addition and multiplication, the order of numbers does matter.
>
> *For example:*
> 6 ÷ 3 is not same as 3 ÷ 6

$$6 ÷ 3 = 2$$
$$3 ÷ 6 = 0.5$$

> We divide **dividend** by **divisor** to get the **quotient.**

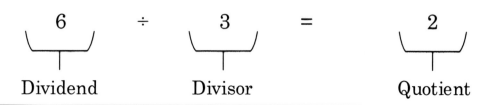

6 ÷ 3 = 2

Dividend   Divisor   Quotient

Division Strategies

 Division is the opposite of multiplication. Multiplication gives us a quick way of doing multiple additions and division gives us a quick way of doing multiple subtractions.

For instance, if you have 6 candies and you eat 2 candies every day. In how many days you will run out of candies?

This can be solved by doing a series of subtractions
- On day 1 you start with 6 candies and end with 6-2 = 4 candies
- On day 2 you will have 4-2 = 2 candies
- On day 3 you will have 2-2 = 0 candies

So answer is 3 days. We could have solved it by doing 6/2 = 3 days.

 When you have to divide, think of multiplication table to get answer quickly.

For instance, if you need to divide 8 by 2 then think of multiplication table of 2 and see how many times 2 gets you 8

2 X				
2	×	1	=	2
2	×	2	=	4
2	×	3	=	6
2	×	4	=	8
2	×	5	=	10
2	×	6	=	12
2	×	7	=	14
2	×	8	=	16
2	×	9	=	18
2	×	10	=	20

Start here

Get result here

 Sometimes we cannot divide things up in equal parts and there may be something left over. For example, there are 5 candies and 2 kids, each gets 2 equal parts and there will be 1 left over. We call this left over as **remainder**

$$5 \div 2 = 2 \text{ R } 1$$

Review: Multiplication Table Reference

1 X

1	×	1	=	1
1	×	2	=	2
1	×	3	=	3
1	×	4	=	4
1	×	5	=	5
1	×	6	=	6
1	×	7	=	7
1	×	8	=	8
1	×	9	=	9
1	×	10	=	10

2 X

2	×	1	=	2
2	×	2	=	4
2	×	3	=	6
2	×	4	=	8
2	×	5	=	10
2	×	6	=	12
2	×	7	=	14
2	×	8	=	16
2	×	9	=	18
2	×	10	=	20

3 X

3	×	1	=	3
3	×	2	=	6
3	×	3	=	9
3	×	4	=	12
3	×	5	=	15
3	×	6	=	18
3	×	7	=	21
3	×	8	=	24
3	×	9	=	27
3	×	10	=	30

4 X

4	×	1	=	4
4	×	2	=	8
4	×	3	=	12
4	×	4	=	16
4	×	5	=	20
4	×	6	=	24
4	×	7	=	28
4	×	8	=	32
4	×	9	=	36
4	×	10	=	40

5 X

5	×	1	=	5
5	×	2	=	10
5	×	3	=	15
5	×	4	=	20
5	×	5	=	25
5	×	6	=	30
5	×	7	=	35
5	×	8	=	40
5	×	9	=	45
5	×	10	=	50

6 X

6	×	1	=	6
6	×	2	=	12
6	×	3	=	18
6	×	4	=	24
6	×	5	=	30
6	×	6	=	36
6	×	7	=	42
6	×	8	=	48
6	×	9	=	54
6	×	10	=	60

7 X

7	×	1	=	7
7	×	2	=	14
7	×	3	=	21
7	×	4	=	28
7	×	5	=	35
7	×	6	=	42
7	×	7	=	49
7	×	8	=	56
7	×	9	=	63
7	×	10	=	70

8 X

8	×	1	=	8
8	×	2	=	16
8	×	3	=	24
8	×	4	=	32
8	×	5	=	40
8	×	6	=	48
8	×	7	=	56
8	×	8	=	64
8	×	9	=	72
8	×	10	=	80

9 X

9	×	1	=	9
9	×	2	=	18
9	×	3	=	27
9	×	4	=	36
9	×	5	=	45
9	×	6	=	54
9	×	7	=	63
9	×	8	=	72
9	×	9	=	81
9	×	10	=	90

10 X

10	×	1	=	10
10	×	2	=	20
10	×	3	=	30
10	×	4	=	40
10	×	5	=	50
10	×	6	=	60
10	×	7	=	70
10	×	8	=	80
10	×	9	=	90
10	×	10	=	100

11 X

11	×	1	=	11
11	×	2	=	22
11	×	3	=	33
11	×	4	=	44
11	×	5	=	55
11	×	6	=	66
11	×	7	=	77
11	×	8	=	88
11	×	9	=	99
11	×	10	=	110

12 X

12	×	1	=	12
12	×	2	=	24
12	×	3	=	36
12	×	4	=	48
12	×	5	=	60
12	×	6	=	72
12	×	7	=	84
12	×	8	=	96
12	×	9	=	108
12	×	10	=	120

Other books from the author that might interest you:

BEGINNERS WORKBOOKS

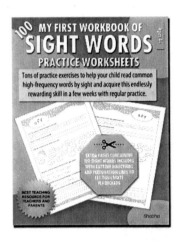

TIME TELLING WORKBOOKS

BASIC MATH FACTS WORKBOOKS

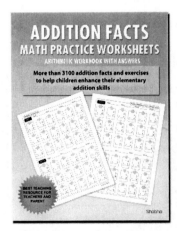
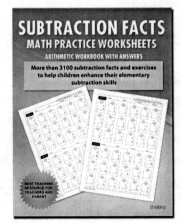
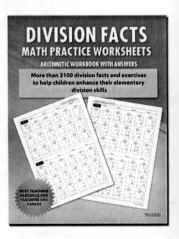

SET I Date: _____ Start: _____ Finish: _____ Score: _____

1
$$2 \div 1$$

2
$$4 \div 1$$

3
$$1 \div 1$$

4
$$10 \div 1$$

5
$$6 \div 1$$

6
$$9 \div 1$$

7
$$7 \div 1$$

8
$$3 \div 1$$

9
$$8 \div 1$$

10
$$5 \div 1$$

11
$$1 \div 1$$

12
$$4 \div 1$$

13
$$8 \div 1$$

14
$$2 \div 1$$

15
$$10 \div 1$$

16
$$7 \div 1$$

17
$$9 \div 1$$

18
$$6 \div 1$$

SET II Date: _____ Start: _____ Finish: _____ Score: _____

19
$$5 \div 1$$

20
$$3 \div 1$$

21
$$6 \div 1$$

22
$$10 \div 1$$

23
$$2 \div 1$$

24
$$5 \div 1$$

25
$$3 \div 1$$

26
$$9 \div 1$$

27
$$7 \div 1$$

28
$$4 \div 1$$

29
$$1 \div 1$$

30
$$8 \div 1$$

31
$$5 \div 1$$

32
$$8 \div 1$$

33
$$2 \div 1$$

34
$$10 \div 1$$

35
$$7 \div 1$$

36
$$4 \div 1$$

SET I Date: _____ Start: _____ Finish: _____ Score: _____

1.
$$9 \div 1$$

2.
$$6 \div 1$$

3.
$$1 \div 1$$

4.
$$7 \div 1$$

5.
$$5 \div 1$$

6.
$$4 \div 1$$

7.
$$8 \div 1$$

8.
$$2 \div 1$$

9.
$$3 \div 1$$

10.
$$1\,0 \div 1$$

11.
$$1 \div 1$$

12.
$$7 \div 1$$

13.
$$1\,0 \div 1$$

14.
$$9 \div 1$$

15.
$$6 \div 1$$

16.
$$5 \div 1$$

17.
$$8 \div 1$$

18.
$$2 \div 1$$

SET II Date: _____ Start: _____ Finish: _____ Score: _____

19.
$$4 \div 1$$

20.
$$3 \div 1$$

21.
$$1 \div 1$$

22.
$$1\,0 \div 1$$

23.
$$2 \div 1$$

24.
$$5 \div 1$$

25.
$$7 \div 1$$

26.
$$6 \div 1$$

27.
$$3 \div 1$$

28.
$$4 \div 1$$

29.
$$8 \div 1$$

30.
$$9 \div 1$$

31.
$$3 \div 1$$

32.
$$5 \div 1$$

33.
$$9 \div 1$$

34.
$$1\,0 \div 1$$

35.
$$8 \div 1$$

36.
$$4 \div 1$$

SET I   Date: _____   Start: _____   Finish: _____   Score: _____

1	2	3	4	5	6
2 0 ÷ 2	4 ÷ 2	1 2 ÷ 2	2 ÷ 2	1 6 ÷ 2	1 0 ÷ 2

7	8	9	10	11	12
1 4 ÷ 2	8 ÷ 2	1 8 ÷ 2	6 ÷ 2	8 ÷ 2	2 0 ÷ 2

13	14	15	16	17	18
2 ÷ 2	4 ÷ 2	1 0 ÷ 2	1 2 ÷ 2	1 8 ÷ 2	6 ÷ 2

SET II   Date: _____   Start: _____   Finish: _____   Score: _____

19	20	21	22	23	24
1 4 ÷ 2	1 6 ÷ 2	2 0 ÷ 2	1 6 ÷ 2	1 2 ÷ 2	1 4 ÷ 2

25	26	27	28	29	30
2 ÷ 2	4 ÷ 2	1 8 ÷ 2	8 ÷ 2	6 ÷ 2	1 0 ÷ 2

31	32	33	34	35	36
1 8 ÷ 2	1 0 ÷ 2	1 2 ÷ 2	2 ÷ 2	4 ÷ 2	1 6 ÷ 2

SET I Date: _____ Start: _____ Finish: _____ Score: _____

1
$$1\,0 \div 2$$

2
$$1\,6 \div 2$$

3
$$6 \div 2$$

4
$$2 \div 2$$

5
$$1\,4 \div 2$$

6
$$1\,2 \div 2$$

7
$$1\,8 \div 2$$

8
$$2\,0 \div 2$$

9
$$8 \div 2$$

10
$$4 \div 2$$

11
$$2\,0 \div 2$$

12
$$1\,8 \div 2$$

13
$$1\,6 \div 2$$

14
$$8 \div 2$$

15
$$1\,2 \div 2$$

16
$$1\,4 \div 2$$

17
$$4 \div 2$$

18
$$1\,0 \div 2$$

SET II Date: _____ Start: _____ Finish: _____ Score: _____

19
$$2 \div 2$$

20
$$6 \div 2$$

21
$$8 \div 2$$

22
$$6 \div 2$$

23
$$1\,0 \div 2$$

24
$$2 \div 2$$

25
$$1\,4 \div 2$$

26
$$1\,6 \div 2$$

27
$$2\,0 \div 2$$

28
$$1\,8 \div 2$$

29
$$4 \div 2$$

30
$$1\,2 \div 2$$

31
$$2 \div 2$$

32
$$1\,2 \div 2$$

33
$$1\,6 \div 2$$

34
$$2\,0 \div 2$$

35
$$1\,4 \div 2$$

36
$$1\,8 \div 2$$

Division Facts

SET I   Date: _____   Start: _____   Finish: _____   Score: _____

1
$$1\ 8$$
$$\div\quad 2$$

2
$$4$$
$$\div\quad 2$$

3
$$2$$
$$\div\quad 2$$

4
$$1\ 4$$
$$\div\quad 2$$

5
$$8$$
$$\div\quad 2$$

6
$$1\ 6$$
$$\div\quad 2$$

7
$$2\ 0$$
$$\div\quad 2$$

8
$$1\ 0$$
$$\div\quad 2$$

9
$$6$$
$$\div\quad 2$$

10
$$1\ 2$$
$$\div\quad 2$$

11
$$4$$
$$\div\quad 2$$

12
$$1\ 0$$
$$\div\quad 2$$

13
$$8$$
$$\div\quad 2$$

14
$$2$$
$$\div\quad 2$$

15
$$1\ 2$$
$$\div\quad 2$$

16
$$2\ 0$$
$$\div\quad 2$$

17
$$1\ 8$$
$$\div\quad 2$$

18
$$6$$
$$\div\quad 2$$

SET II   Date: _____   Start: _____   Finish: _____   Score: _____

19
$$1\ 6$$
$$\div\quad 2$$

20
$$1\ 4$$
$$\div\quad 2$$

21
$$2$$
$$\div\quad 2$$

22
$$1\ 8$$
$$\div\quad 2$$

23
$$4$$
$$\div\quad 2$$

24
$$1\ 4$$
$$\div\quad 2$$

25
$$1\ 6$$
$$\div\quad 2$$

26
$$8$$
$$\div\quad 2$$

27
$$1\ 2$$
$$\div\quad 2$$

28
$$6$$
$$\div\quad 2$$

29
$$1\ 0$$
$$\div\quad 2$$

30
$$2\ 0$$
$$\div\quad 2$$

31
$$4$$
$$\div\quad 2$$

32
$$1\ 0$$
$$\div\quad 2$$

33
$$1\ 4$$
$$\div\quad 2$$

34
$$1\ 2$$
$$\div\quad 2$$

35
$$2\ 0$$
$$\div\quad 2$$

36
$$8$$
$$\div\quad 2$$

SET I Date: _____ Start: _____ Finish: _____ Score: _____

1	2	3	4	5	6
$4 \div 2$	$10 \div 2$	$6 \div 2$	$14 \div 2$	$18 \div 2$	$2 \div 2$

7	8	9	10	11	12
$20 \div 2$	$16 \div 2$	$12 \div 2$	$8 \div 2$	$16 \div 2$	$10 \div 2$

13	14	15	16	17	18
$2 \div 2$	$18 \div 2$	$6 \div 2$	$20 \div 2$	$8 \div 2$	$4 \div 2$

SET II Date: _____ Start: _____ Finish: _____ Score: _____

19	20	21	22	23	24
$12 \div 2$	$14 \div 2$	$2 \div 2$	$20 \div 2$	$4 \div 2$	$14 \div 2$

25	26	27	28	29	30
$18 \div 2$	$16 \div 2$	$10 \div 2$	$8 \div 2$	$12 \div 2$	$6 \div 2$

31	32	33	34	35	36
$20 \div 2$	$6 \div 2$	$2 \div 2$	$12 \div 2$	$14 \div 2$	$18 \div 2$

SET I Date: _____ Start: _____ Finish: _____ Score: _____

1) 3 0 ÷ 3

2) 1 2 ÷ 3

3) 6 ÷ 3

4) 2 4 ÷ 3

5) 2 1 ÷ 3

6) 1 5 ÷ 3

7) 1 8 ÷ 3

8) 2 7 ÷ 3

9) 3 ÷ 3

10) 9 ÷ 3

11) 1 2 ÷ 3

12) 2 7 ÷ 3

13) 1 5 ÷ 3

14) 2 1 ÷ 3

15) 3 0 ÷ 3

16) 9 ÷ 3

17) 6 ÷ 3

18) 2 4 ÷ 3

SET II Date: _____ Start: _____ Finish: _____ Score: _____

19) 3 ÷ 3

20) 1 8 ÷ 3

21) 6 ÷ 3

22) 1 5 ÷ 3

23) 3 0 ÷ 3

24) 3 ÷ 3

25) 9 ÷ 3

26) 2 4 ÷ 3

27) 2 7 ÷ 3

28) 1 8 ÷ 3

29) 1 2 ÷ 3

30) 2 1 ÷ 3

31) 6 ÷ 3

32) 1 2 ÷ 3

33) 3 ÷ 3

34) 2 7 ÷ 3

35) 3 0 ÷ 3

36) 2 1 ÷ 3

SET I Date: _____ Start: _____ Finish: _____ Score: _____

1	2	3	4	5	6
6 $\div\ 3$	$1\ 5$ $\div\ 3$	$2\ 7$ $\div\ 3$	9 $\div\ 3$	$1\ 2$ $\div\ 3$	$2\ 1$ $\div\ 3$

7	8	9	10	11	12
$3\ 0$ $\div\ 3$	3 $\div\ 3$	$1\ 8$ $\div\ 3$	$2\ 4$ $\div\ 3$	$2\ 7$ $\div\ 3$	6 $\div\ 3$

13	14	15	16	17	18
$1\ 2$ $\div\ 3$	$1\ 5$ $\div\ 3$	$2\ 1$ $\div\ 3$	$2\ 4$ $\div\ 3$	$1\ 8$ $\div\ 3$	$3\ 0$ $\div\ 3$

SET II Date: _____ Start: _____ Finish: _____ Score: _____

19	20	21	22	23	24
9 $\div\ 3$	3 $\div\ 3$	$2\ 4$ $\div\ 3$	$2\ 1$ $\div\ 3$	9 $\div\ 3$	$1\ 5$ $\div\ 3$

25	26	27	28	29	30
$1\ 2$ $\div\ 3$	3 $\div\ 3$	6 $\div\ 3$	$3\ 0$ $\div\ 3$	$1\ 8$ $\div\ 3$	$2\ 7$ $\div\ 3$

31	32	33	34	35	36
6 $\div\ 3$	3 $\div\ 3$	$3\ 0$ $\div\ 3$	$2\ 1$ $\div\ 3$	$1\ 8$ $\div\ 3$	$2\ 7$ $\div\ 3$

SET I   Date: _____   Start: _____   Finish: _____   Score: _____

1	**2**	**3**	**4**	**5**	**6**
1 5 ÷ 3	3 ÷ 3	3 0 ÷ 3	6 ÷ 3	9 ÷ 3	1 8 ÷ 3
7	**8**	**9**	**10**	**11**	**12**
1 2 ÷ 3	2 4 ÷ 3	2 7 ÷ 3	2 1 ÷ 3	1 8 ÷ 3	2 7 ÷ 3
13	**14**	**15**	**16**	**17**	**18**
2 1 ÷ 3	1 5 ÷ 3	3 ÷ 3	2 4 ÷ 3	6 ÷ 3	1 2 ÷ 3

SET II   Date: _____   Start: _____   Finish: _____   Score: _____

19	**20**	**21**	**22**	**23**	**24**
9 ÷ 3	3 0 ÷ 3	3 ÷ 3	1 5 ÷ 3	9 ÷ 3	1 2 ÷ 3
25	**26**	**27**	**28**	**29**	**30**
2 7 ÷ 3	1 8 ÷ 3	2 4 ÷ 3	3 0 ÷ 3	6 ÷ 3	2 1 ÷ 3
31	**32**	**33**	**34**	**35**	**36**
2 4 ÷ 3	6 ÷ 3	2 7 ÷ 3	3 0 ÷ 3	9 ÷ 3	1 8 ÷ 3

SET I Date: _____ Start: _____ Finish: _____ Score: _____

1
$$21 \div 3$$

2
$$6 \div 3$$

3
$$3 \div 3$$

4
$$15 \div 3$$

5
$$9 \div 3$$

6
$$30 \div 3$$

7
$$18 \div 3$$

8
$$27 \div 3$$

9
$$12 \div 3$$

10
$$24 \div 3$$

11
$$9 \div 3$$

12
$$6 \div 3$$

13
$$21 \div 3$$

14
$$3 \div 3$$

15
$$18 \div 3$$

16
$$24 \div 3$$

17
$$12 \div 3$$

18
$$30 \div 3$$

SET II Date: _____ Start: _____ Finish: _____ Score: _____

19
$$27 \div 3$$

20
$$15 \div 3$$

21
$$21 \div 3$$

22
$$9 \div 3$$

23
$$3 \div 3$$

24
$$6 \div 3$$

25
$$12 \div 3$$

26
$$27 \div 3$$

27
$$15 \div 3$$

28
$$24 \div 3$$

29
$$30 \div 3$$

30
$$18 \div 3$$

31
$$6 \div 3$$

32
$$15 \div 3$$

33
$$18 \div 3$$

34
$$9 \div 3$$

35
$$27 \div 3$$

36
$$24 \div 3$$

SET I   Date: _____   Start: _____   Finish: _____   Score: _____

1)
$$4 \div 2$$

2)
$$1\ 4 \div 2$$

3)
$$6 \div 3$$

4)
$$1\ 5 \div 3$$

5)
$$2\ 7 \div 3$$

6)
$$6 \div 2$$

7)
$$2\ 1 \div 3$$

8)
$$3 \div 3$$

9)
$$2\ 4 \div 3$$

10)
$$9 \div 3$$

11)
$$3\ 0 \div 3$$

12)
$$1\ 8 \div 3$$

13)
$$1\ 2 \div 2$$

14)
$$2 \div 2$$

15)
$$1\ 2 \div 3$$

16)
$$1\ 6 \div 2$$

17)
$$8 \div 2$$

18)
$$1\ 0 \div 2$$

SET II   Date: _____   Start: _____   Finish: _____   Score: _____

19)
$$2\ 0 \div 2$$

20)
$$1\ 8 \div 2$$

21)
$$4 \div 2$$

22)
$$1\ 4 \div 2$$

23)
$$6 \div 3$$

24)
$$1\ 5 \div 3$$

25)
$$2\ 7 \div 3$$

26)
$$6 \div 2$$

27)
$$2\ 1 \div 3$$

28)
$$3 \div 3$$

29)
$$2\ 4 \div 3$$

30)
$$9 \div 3$$

31)
$$3\ 0 \div 3$$

32)
$$1\ 8 \div 3$$

33)
$$1\ 2 \div 2$$

34)
$$2 \div 2$$

35)
$$1\ 2 \div 3$$

36)
$$1\ 6 \div 2$$

SET I Date: _____ Start: _____ Finish: _____ Score: _____

1
$$8 \div 2$$

2
$$1\ 4 \div 2$$

3
$$1\ 8 \div 3$$

4
$$3\ 0 \div 3$$

5
$$1\ 2 \div 3$$

6
$$6 \div 2$$

7
$$1\ 6 \div 2$$

8
$$2\ 4 \div 3$$

9
$$2\ 0 \div 2$$

10
$$1\ 0 \div 2$$

11
$$1\ 8 \div 2$$

12
$$2\ 7 \div 3$$

13
$$1\ 5 \div 3$$

14
$$6 \div 3$$

15
$$2\ 1 \div 3$$

16
$$3 \div 3$$

17
$$1\ 2 \div 2$$

18
$$4 \div 2$$

SET II Date: _____ Start: _____ Finish: _____ Score: _____

19
$$2 \div 2$$

20
$$9 \div 3$$

21
$$8 \div 2$$

22
$$1\ 4 \div 2$$

23
$$1\ 8 \div 3$$

24
$$3\ 0 \div 3$$

25
$$1\ 2 \div 3$$

26
$$6 \div 2$$

27
$$1\ 6 \div 2$$

28
$$2\ 4 \div 3$$

29
$$2\ 0 \div 2$$

30
$$1\ 0 \div 2$$

31
$$1\ 8 \div 2$$

32
$$2\ 7 \div 3$$

33
$$1\ 5 \div 3$$

34
$$6 \div 3$$

35
$$2\ 1 \div 3$$

36
$$3 \div 3$$

SET I   Date:_____   Start:_____   Finish:_____   Score:_____

1
$$16 \div 2$$

2
$$20 \div 2$$

3
$$6 \div 2$$

4
$$15 \div 3$$

5
$$2 \div 2$$

6
$$6 \div 3$$

7
$$4 \div 2$$

8
$$3 \div 3$$

9
$$10 \div 2$$

10
$$27 \div 3$$

11
$$30 \div 3$$

12
$$18 \div 3$$

13
$$8 \div 2$$

14
$$24 \div 3$$

15
$$14 \div 2$$

16
$$9 \div 3$$

17
$$21 \div 3$$

18
$$18 \div 2$$

SET II   Date:_____   Start:_____   Finish:_____   Score:_____

19
$$12 \div 3$$

20
$$12 \div 2$$

21
$$16 \div 2$$

22
$$20 \div 2$$

23
$$6 \div 2$$

24
$$15 \div 3$$

25
$$2 \div 2$$

26
$$6 \div 3$$

27
$$4 \div 2$$

28
$$3 \div 3$$

29
$$10 \div 2$$

30
$$27 \div 3$$

31
$$30 \div 3$$

32
$$18 \div 3$$

33
$$8 \div 2$$

34
$$24 \div 3$$

35
$$14 \div 2$$

36
$$9 \div 3$$

Division Facts

SET I

Date: _____ Start: _____ Finish: _____ Score: _____

1
$$18 \div 3$$

2
$$3 \div 3$$

3
$$16 \div 2$$

4
$$12 \div 2$$

5
$$9 \div 3$$

6
$$21 \div 3$$

7
$$27 \div 3$$

8
$$24 \div 3$$

9
$$6 \div 3$$

10
$$30 \div 3$$

11
$$12 \div 3$$

12
$$10 \div 2$$

13
$$14 \div 2$$

14
$$8 \div 2$$

15
$$18 \div 2$$

16
$$20 \div 2$$

17
$$15 \div 3$$

18
$$2 \div 2$$

SET II

Date: _____ Start: _____ Finish: _____ Score: _____

19
$$6 \div 2$$

20
$$4 \div 2$$

21
$$18 \div 3$$

22
$$3 \div 3$$

23
$$16 \div 2$$

24
$$12 \div 2$$

25
$$9 \div 3$$

26
$$21 \div 3$$

27
$$27 \div 3$$

28
$$24 \div 3$$

29
$$6 \div 3$$

30
$$30 \div 3$$

31
$$12 \div 3$$

32
$$10 \div 2$$

33
$$14 \div 2$$

34
$$8 \div 2$$

35
$$18 \div 2$$

36
$$20 \div 2$$

Division Facts

SET I

Date: _____   Start: _____   Finish: _____   Score: _____

1	2	3	4	5	6
1 6 ÷ 2	8 ÷ 2	1 4 ÷ 2	1 5 ÷ 3	2 0 ÷ 2	3 0 ÷ 3

7	8	9	10	11	12
2 1 ÷ 3	1 0 ÷ 2	1 2 ÷ 2	3 ÷ 3	1 8 ÷ 3	6 ÷ 3

13	14	15	16	17	18
2 4 ÷ 3	1 2 ÷ 3	2 7 ÷ 3	6 ÷ 2	4 ÷ 2	9 ÷ 3

SET II

Date: _____   Start: _____   Finish: _____   Score: _____

19	20	21	22	23	24
2 ÷ 2	1 8 ÷ 2	1 6 ÷ 2	8 ÷ 2	1 4 ÷ 2	1 5 ÷ 3

25	26	27	28	29	30
2 0 ÷ 2	3 0 ÷ 3	2 1 ÷ 3	1 0 ÷ 2	1 2 ÷ 2	3 ÷ 3

31	32	33	34	35	36
1 8 ÷ 3	6 ÷ 3	2 4 ÷ 3	1 2 ÷ 3	2 7 ÷ 3	6 ÷ 2

SET I Date: _____ Start: _____ Finish: _____ Score: _____

1
$$24 \div 3$$

2
$$2 \div 2$$

3
$$16 \div 2$$

4
$$12 \div 3$$

5
$$12 \div 2$$

6
$$21 \div 3$$

7
$$3 \div 3$$

8
$$14 \div 2$$

9
$$9 \div 3$$

10
$$18 \div 3$$

11
$$6 \div 3$$

12
$$4 \div 2$$

13
$$8 \div 2$$

14
$$15 \div 3$$

15
$$10 \div 2$$

16
$$18 \div 2$$

17
$$20 \div 2$$

18
$$6 \div 2$$

SET II Date: _____ Start: _____ Finish: _____ Score: _____

19
$$30 \div 3$$

20
$$27 \div 3$$

21
$$24 \div 3$$

22
$$2 \div 2$$

23
$$16 \div 2$$

24
$$12 \div 3$$

25
$$12 \div 2$$

26
$$21 \div 3$$

27
$$3 \div 3$$

28
$$14 \div 2$$

29
$$9 \div 3$$

30
$$18 \div 3$$

31
$$6 \div 3$$

32
$$4 \div 2$$

33
$$8 \div 2$$

34
$$15 \div 3$$

35
$$10 \div 2$$

36
$$18 \div 2$$

SET I   Date: _____   Start: _____   Finish: _____   Score: _____

1	2	3	4	5	6
3 2 ÷ 4	8 ÷ 4	2 8 ÷ 4	4 0 ÷ 4	2 0 ÷ 4	3 6 ÷ 4

7	8	9	10	11	12
1 6 ÷ 4	4 ÷ 4	1 2 ÷ 4	2 4 ÷ 4	2 8 ÷ 4	2 4 ÷ 4

13	14	15	16	17	18
3 2 ÷ 4	1 6 ÷ 4	4 0 ÷ 4	1 2 ÷ 4	8 ÷ 4	4 ÷ 4

SET II   Date: _____   Start: _____   Finish: _____   Score: _____

19	20	21	22	23	24
2 0 ÷ 4	3 6 ÷ 4	1 6 ÷ 4	3 2 ÷ 4	2 4 ÷ 4	2 8 ÷ 4

25	26	27	28	29	30
8 ÷ 4	4 ÷ 4	3 6 ÷ 4	1 2 ÷ 4	2 0 ÷ 4	4 0 ÷ 4

31	32	33	34	35	36
2 0 ÷ 4	8 ÷ 4	2 4 ÷ 4	2 8 ÷ 4	4 ÷ 4	1 2 ÷ 4

SET I Date: _____ Start: _____ Finish: _____ Score: _____

1	2	3	4	5	6
3 2 ÷ 4	1 6 ÷ 4	2 4 ÷ 4	3 6 ÷ 4	2 0 ÷ 4	1 2 ÷ 4

7	8	9	10	11	12
4 0 ÷ 4	8 ÷ 4	2 8 ÷ 4	4 ÷ 4	3 6 ÷ 4	1 2 ÷ 4

13	14	15	16	17	18
1 6 ÷ 4	4 0 ÷ 4	2 0 ÷ 4	4 ÷ 4	3 2 ÷ 4	2 4 ÷ 4

SET II Date: _____ Start: _____ Finish: _____ Score: _____

19	20	21	22	23	24
2 8 ÷ 4	8 ÷ 4	3 2 ÷ 4	4 ÷ 4	1 6 ÷ 4	2 8 ÷ 4

25	26	27	28	29	30
4 0 ÷ 4	1 2 ÷ 4	2 0 ÷ 4	2 4 ÷ 4	8 ÷ 4	3 6 ÷ 4

31	32	33	34	35	36
1 6 ÷ 4	4 0 ÷ 4	1 2 ÷ 4	2 0 ÷ 4	4 ÷ 4	3 2 ÷ 4

SET I   Date: _____   Start: _____   Finish: _____   Score: _____

1

$$3\ 6 \div \quad 4$$

2

$$2\ 4 \div \quad 4$$

3

$$4 \div \quad 4$$

4

$$2\ 8 \div \quad 4$$

5

$$1\ 6 \div \quad 4$$

6

$$1\ 2 \div \quad 4$$

7

$$4\ 0 \div \quad 4$$

8

$$3\ 2 \div \quad 4$$

9

$$8 \div \quad 4$$

10

$$2\ 0 \div \quad 4$$

11

$$8 \div \quad 4$$

12

$$2\ 0 \div \quad 4$$

13

$$2\ 4 \div \quad 4$$

14

$$1\ 2 \div \quad 4$$

15

$$4\ 0 \div \quad 4$$

16

$$4 \div \quad 4$$

17

$$3\ 6 \div \quad 4$$

18

$$3\ 2 \div \quad 4$$

SET II   Date: _____   Start: _____   Finish: _____   Score: _____

19

$$1\ 6 \div \quad 4$$

20

$$2\ 8 \div \quad 4$$

21

$$3\ 2 \div \quad 4$$

22

$$1\ 6 \div \quad 4$$

23

$$2\ 0 \div \quad 4$$

24

$$8 \div \quad 4$$

25

$$3\ 6 \div \quad 4$$

26

$$4 \div \quad 4$$

27

$$1\ 2 \div \quad 4$$

28

$$2\ 4 \div \quad 4$$

29

$$4\ 0 \div \quad 4$$

30

$$2\ 8 \div \quad 4$$

31

$$3\ 6 \div \quad 4$$

32

$$8 \div \quad 4$$

33

$$4\ 0 \div \quad 4$$

34

$$4 \div \quad 4$$

35

$$1\ 6 \div \quad 4$$

36

$$2\ 4 \div \quad 4$$

SET I Date: _____ Start: _____ Finish: _____ Score: _____

1	**2**	**3**	**4**	**5**	**6**
3 2 ÷ 4	4 ÷ 4	3 6 ÷ 4	2 8 ÷ 4	8 ÷ 4	2 0 ÷ 4
7	**8**	**9**	**10**	**11**	**12**
1 2 ÷ 4	4 0 ÷ 4	1 6 ÷ 4	2 4 ÷ 4	3 2 ÷ 4	1 2 ÷ 4
13	**14**	**15**	**16**	**17**	**18**
4 0 ÷ 4	8 ÷ 4	3 6 ÷ 4	2 4 ÷ 4	4 ÷ 4	2 8 ÷ 4

SET II Date: _____ Start: _____ Finish: _____ Score: _____

19	**20**	**21**	**22**	**23**	**24**
2 0 ÷ 4	1 6 ÷ 4	3 6 ÷ 4	4 ÷ 4	1 6 ÷ 4	2 4 ÷ 4
25	**26**	**27**	**28**	**29**	**30**
3 2 ÷ 4	4 0 ÷ 4	2 8 ÷ 4	1 2 ÷ 4	2 0 ÷ 4	8 ÷ 4
31	**32**	**33**	**34**	**35**	**36**
4 ÷ 4	2 0 ÷ 4	3 6 ÷ 4	2 8 ÷ 4	1 2 ÷ 4	4 0 ÷ 4

SET I   Date: _____   Start: _____   Finish: _____   Score: _____

(1) 1 5 ÷ 5	(2) 2 0 ÷ 5	(3) 2 5 ÷ 5	(4) 1 0 ÷ 5	(5) 4 0 ÷ 5	(6) 4 5 ÷ 5
(7) 5 ÷ 5	(8) 3 0 ÷ 5	(9) 3 5 ÷ 5	(10) 5 0 ÷ 5	(11) 4 0 ÷ 5	(12) 2 0 ÷ 5
(13) 1 0 ÷ 5	(14) 4 5 ÷ 5	(15) 1 5 ÷ 5	(16) 3 5 ÷ 5	(17) 3 0 ÷ 5	(18) 2 5 ÷ 5

SET II   Date: _____   Start: _____   Finish: _____   Score: _____

(19) 5 0 ÷ 5	(20) 5 ÷ 5	(21) 1 0 ÷ 5	(22) 3 5 ÷ 5	(23) 5 ÷ 5	(24) 2 0 ÷ 5
(25) 3 0 ÷ 5	(26) 5 0 ÷ 5	(27) 1 5 ÷ 5	(28) 2 5 ÷ 5	(29) 4 5 ÷ 5	(30) 4 0 ÷ 5
(31) 3 0 ÷ 5	(32) 1 0 ÷ 5	(33) 5 ÷ 5	(34) 1 5 ÷ 5	(35) 2 0 ÷ 5	(36) 4 0 ÷ 5

SET I   Date: _____   Start: _____   Finish: _____   Score: _____

1

$$
\begin{array}{r}
2\ 0 \\
\div\quad 5 \\
\hline
\end{array}
$$

2

$$
\begin{array}{r}
4\ 5 \\
\div\quad 5 \\
\hline
\end{array}
$$

3

$$
\begin{array}{r}
3\ 0 \\
\div\quad 5 \\
\hline
\end{array}
$$

4

$$
\begin{array}{r}
1\ 0 \\
\div\quad 5 \\
\hline
\end{array}
$$

5

$$
\begin{array}{r}
3\ 5 \\
\div\quad 5 \\
\hline
\end{array}
$$

6

$$
\begin{array}{r}
2\ 5 \\
\div\quad 5 \\
\hline
\end{array}
$$

7

$$
\begin{array}{r}
5 \\
\div\quad 5 \\
\hline
\end{array}
$$

8

$$
\begin{array}{r}
4\ 0 \\
\div\quad 5 \\
\hline
\end{array}
$$

9

$$
\begin{array}{r}
1\ 5 \\
\div\quad 5 \\
\hline
\end{array}
$$

10

$$
\begin{array}{r}
5\ 0 \\
\div\quad 5 \\
\hline
\end{array}
$$

11

$$
\begin{array}{r}
1\ 0 \\
\div\quad 5 \\
\hline
\end{array}
$$

12

$$
\begin{array}{r}
4\ 0 \\
\div\quad 5 \\
\hline
\end{array}
$$

13

$$
\begin{array}{r}
4\ 5 \\
\div\quad 5 \\
\hline
\end{array}
$$

14

$$
\begin{array}{r}
2\ 0 \\
\div\quad 5 \\
\hline
\end{array}
$$

15

$$
\begin{array}{r}
1\ 5 \\
\div\quad 5 \\
\hline
\end{array}
$$

16

$$
\begin{array}{r}
2\ 5 \\
\div\quad 5 \\
\hline
\end{array}
$$

17

$$
\begin{array}{r}
3\ 0 \\
\div\quad 5 \\
\hline
\end{array}
$$

18

$$
\begin{array}{r}
5\ 0 \\
\div\quad 5 \\
\hline
\end{array}
$$

SET II   Date: _____   Start: _____   Finish: _____   Score: _____

19

$$
\begin{array}{r}
3\ 5 \\
\div\quad 5 \\
\hline
\end{array}
$$

20

$$
\begin{array}{r}
5 \\
\div\quad 5 \\
\hline
\end{array}
$$

21

$$
\begin{array}{r}
3\ 5 \\
\div\quad 5 \\
\hline
\end{array}
$$

22

$$
\begin{array}{r}
5 \\
\div\quad 5 \\
\hline
\end{array}
$$

23

$$
\begin{array}{r}
2\ 5 \\
\div\quad 5 \\
\hline
\end{array}
$$

24

$$
\begin{array}{r}
1\ 5 \\
\div\quad 5 \\
\hline
\end{array}
$$

25

$$
\begin{array}{r}
3\ 0 \\
\div\quad 5 \\
\hline
\end{array}
$$

26

$$
\begin{array}{r}
4\ 0 \\
\div\quad 5 \\
\hline
\end{array}
$$

27

$$
\begin{array}{r}
5\ 0 \\
\div\quad 5 \\
\hline
\end{array}
$$

28

$$
\begin{array}{r}
4\ 5 \\
\div\quad 5 \\
\hline
\end{array}
$$

29

$$
\begin{array}{r}
1\ 0 \\
\div\quad 5 \\
\hline
\end{array}
$$

30

$$
\begin{array}{r}
2\ 0 \\
\div\quad 5 \\
\hline
\end{array}
$$

31

$$
\begin{array}{r}
2\ 5 \\
\div\quad 5 \\
\hline
\end{array}
$$

32

$$
\begin{array}{r}
2\ 0 \\
\div\quad 5 \\
\hline
\end{array}
$$

33

$$
\begin{array}{r}
1\ 5 \\
\div\quad 5 \\
\hline
\end{array}
$$

34

$$
\begin{array}{r}
3\ 5 \\
\div\quad 5 \\
\hline
\end{array}
$$

35

$$
\begin{array}{r}
4\ 5 \\
\div\quad 5 \\
\hline
\end{array}
$$

36

$$
\begin{array}{r}
4\ 0 \\
\div\quad 5 \\
\hline
\end{array}
$$

SET I   Date: _____   Start: _____   Finish: _____   Score: _____

(1) $10 \div 5$	(2) $15 \div 5$	(3) $30 \div 5$	(4) $40 \div 5$	(5) $35 \div 5$	(6) $50 \div 5$
(7) $45 \div 5$	(8) $25 \div 5$	(9) $5 \div 5$	(10) $20 \div 5$	(11) $50 \div 5$	(12) $35 \div 5$
(13) $25 \div 5$	(14) $15 \div 5$	(15) $20 \div 5$	(16) $40 \div 5$	(17) $30 \div 5$	(18) $10 \div 5$

SET II   Date: _____   Start: _____   Finish: _____   Score: _____

(19) $5 \div 5$	(20) $45 \div 5$	(21) $35 \div 5$	(22) $40 \div 5$	(23) $5 \div 5$	(24) $10 \div 5$
(25) $15 \div 5$	(26) $50 \div 5$	(27) $45 \div 5$	(28) $30 \div 5$	(29) $20 \div 5$	(30) $25 \div 5$
(31) $50 \div 5$	(32) $5 \div 5$	(33) $40 \div 5$	(34) $45 \div 5$	(35) $15 \div 5$	(36) $20 \div 5$

Division Facts

SET I Date: _____ Start: _____ Finish: _____ Score: _____

1)
$$45 \div 5$$

2)
$$35 \div 5$$

3)
$$25 \div 5$$

4)
$$10 \div 5$$

5)
$$50 \div 5$$

6)
$$20 \div 5$$

7)
$$40 \div 5$$

8)
$$15 \div 5$$

9)
$$30 \div 5$$

10)
$$5 \div 5$$

11)
$$40 \div 5$$

12)
$$15 \div 5$$

13)
$$10 \div 5$$

14)
$$50 \div 5$$

15)
$$45 \div 5$$

16)
$$35 \div 5$$

17)
$$30 \div 5$$

18)
$$20 \div 5$$

SET II Date: _____ Start: _____ Finish: _____ Score: _____

19)
$$25 \div 5$$

20)
$$5 \div 5$$

21)
$$30 \div 5$$

22)
$$50 \div 5$$

23)
$$15 \div 5$$

24)
$$10 \div 5$$

25)
$$40 \div 5$$

26)
$$35 \div 5$$

27)
$$45 \div 5$$

28)
$$25 \div 5$$

29)
$$5 \div 5$$

30)
$$20 \div 5$$

31)
$$35 \div 5$$

32)
$$5 \div 5$$

33)
$$50 \div 5$$

34)
$$45 \div 5$$

35)
$$20 \div 5$$

36)
$$40 \div 5$$

SET I   Date: _____   Start: _____   Finish: _____   Score: _____

1
$$15 \div 5$$

2
$$35 \div 5$$

3
$$24 \div 4$$

4
$$20 \div 4$$

5
$$32 \div 4$$

6
$$50 \div 5$$

7
$$10 \div 5$$

8
$$12 \div 4$$

9
$$8 \div 4$$

10
$$30 \div 5$$

11
$$5 \div 5$$

12
$$28 \div 4$$

13
$$36 \div 4$$

14
$$4 \div 4$$

15
$$20 \div 5$$

16
$$40 \div 4$$

17
$$25 \div 5$$

18
$$40 \div 5$$

SET II   Date: _____   Start: _____   Finish: _____   Score: _____

19
$$16 \div 4$$

20
$$45 \div 5$$

21
$$15 \div 5$$

22
$$35 \div 5$$

23
$$24 \div 4$$

24
$$20 \div 4$$

25
$$32 \div 4$$

26
$$50 \div 5$$

27
$$10 \div 5$$

28
$$12 \div 4$$

29
$$8 \div 4$$

30
$$30 \div 5$$

31
$$5 \div 5$$

32
$$28 \div 4$$

33
$$36 \div 4$$

34
$$4 \div 4$$

35
$$20 \div 5$$

36
$$40 \div 4$$

Review: Dividing by 4 and 5

SET I Date: _____ Start: _____ Finish: _____ Score: _____

1
$$25 \div 5$$

2
$$35 \div 5$$

3
$$4 \div 4$$

4
$$45 \div 5$$

5
$$30 \div 5$$

6
$$28 \div 4$$

7
$$20 \div 4$$

8
$$8 \div 4$$

9
$$24 \div 4$$

10
$$12 \div 4$$

11
$$40 \div 4$$

12
$$5 \div 5$$

13
$$20 \div 5$$

14
$$10 \div 5$$

15
$$50 \div 5$$

16
$$40 \div 5$$

17
$$16 \div 4$$

18
$$32 \div 4$$

SET II Date: _____ Start: _____ Finish: _____ Score: _____

19
$$36 \div 4$$

20
$$15 \div 5$$

21
$$25 \div 5$$

22
$$35 \div 5$$

23
$$4 \div 4$$

24
$$45 \div 5$$

25
$$30 \div 5$$

26
$$28 \div 4$$

27
$$20 \div 4$$

28
$$8 \div 4$$

29
$$24 \div 4$$

30
$$12 \div 4$$

31
$$40 \div 4$$

32
$$5 \div 5$$

33
$$20 \div 5$$

34
$$10 \div 5$$

35
$$50 \div 5$$

36
$$40 \div 5$$

Division Facts

SET I Date: _____ Start: _____ Finish: _____ Score: _____

1
$30 \div 5$

2
$16 \div 4$

3
$35 \div 5$

4
$8 \div 4$

5
$28 \div 4$

6
$25 \div 5$

7
$40 \div 4$

8
$40 \div 5$

9
$20 \div 4$

10
$4 \div 4$

11
$12 \div 4$

12
$36 \div 4$

13
$24 \div 4$

14
$20 \div 5$

15
$5 \div 5$

16
$45 \div 5$

17
$10 \div 5$

18
$32 \div 4$

SET II Date: _____ Start: _____ Finish: _____ Score: _____

19
$15 \div 5$

20
$50 \div 5$

21
$30 \div 5$

22
$16 \div 4$

23
$35 \div 5$

24
$8 \div 4$

25
$28 \div 4$

26
$25 \div 5$

27
$40 \div 4$

28
$40 \div 5$

29
$20 \div 4$

30
$4 \div 4$

31
$12 \div 4$

32
$36 \div 4$

33
$24 \div 4$

34
$20 \div 5$

35
$5 \div 5$

36
$45 \div 5$

SET I Date: _____ Start: _____ Finish: _____ Score: _____

1.
$$2\ 5 \div 5$$

2.
$$1\ 2 \div 4$$

3.
$$1\ 5 \div 5$$

4.
$$2\ 0 \div 4$$

5.
$$2\ 0 \div 5$$

6.
$$3\ 6 \div 4$$

7.
$$2\ 8 \div 4$$

8.
$$4\ 0 \div 5$$

9.
$$4\ 5 \div 5$$

10.
$$1\ 0 \div 5$$

11.
$$5 \div 5$$

12.
$$3\ 2 \div 4$$

13.
$$5\ 0 \div 5$$

14.
$$4 \div 4$$

15.
$$4\ 0 \div 4$$

16.
$$8 \div 4$$

17.
$$1\ 6 \div 4$$

18.
$$2\ 4 \div 4$$

SET II Date: _____ Start: _____ Finish: _____ Score: _____

19.
$$3\ 0 \div 5$$

20.
$$3\ 5 \div 5$$

21.
$$2\ 5 \div 5$$

22.
$$1\ 2 \div 4$$

23.
$$1\ 5 \div 5$$

24.
$$2\ 0 \div 4$$

25.
$$2\ 0 \div 5$$

26.
$$3\ 6 \div 4$$

27.
$$2\ 8 \div 4$$

28.
$$4\ 0 \div 5$$

29.
$$4\ 5 \div 5$$

30.
$$1\ 0 \div 5$$

31.
$$5 \div 5$$

32.
$$3\ 2 \div 4$$

33.
$$5\ 0 \div 5$$

34.
$$4 \div 4$$

35.
$$4\ 0 \div 4$$

36.
$$8 \div 4$$

SET I Date: _____   Start: _____   Finish: _____   Score: _____

1
$$\begin{array}{r} 1\ 6 \\ \div\quad 2 \\ \hline \end{array}$$

2
$$\begin{array}{r} 2\ 0 \\ \div\quad 2 \\ \hline \end{array}$$

3
$$\begin{array}{r} 3\ 6 \\ \div\quad 4 \\ \hline \end{array}$$

4
$$\begin{array}{r} 1\ 8 \\ \div\quad 2 \\ \hline \end{array}$$

5
$$\begin{array}{r} 1\ 6 \\ \div\quad 4 \\ \hline \end{array}$$

6
$$\begin{array}{r} 2\ 7 \\ \div\quad 3 \\ \hline \end{array}$$

7
$$\begin{array}{r} 8 \\ \div\quad 4 \\ \hline \end{array}$$

8
$$\begin{array}{r} 2\ 8 \\ \div\quad 4 \\ \hline \end{array}$$

9
$$\begin{array}{r} 2\ 5 \\ \div\quad 5 \\ \hline \end{array}$$

10
$$\begin{array}{r} 8 \\ \div\quad 2 \\ \hline \end{array}$$

11
$$\begin{array}{r} 1\ 2 \\ \div\quad 3 \\ \hline \end{array}$$

12
$$\begin{array}{r} 3\ 2 \\ \div\quad 4 \\ \hline \end{array}$$

13
$$\begin{array}{r} 4\ 0 \\ \div\quad 5 \\ \hline \end{array}$$

14
$$\begin{array}{r} 2\ 0 \\ \div\quad 4 \\ \hline \end{array}$$

15
$$\begin{array}{r} 1\ 8 \\ \div\quad 3 \\ \hline \end{array}$$

16
$$\begin{array}{r} 2\ 4 \\ \div\quad 3 \\ \hline \end{array}$$

17
$$\begin{array}{r} 5\ 0 \\ \div\quad 5 \\ \hline \end{array}$$

18
$$\begin{array}{r} 3\ 5 \\ \div\quad 5 \\ \hline \end{array}$$

SET II Date: _____   Start: _____   Finish: _____   Score: _____

19
$$\begin{array}{r} 1\ 5 \\ \div\quad 5 \\ \hline \end{array}$$

20
$$\begin{array}{r} 3\ 0 \\ \div\quad 3 \\ \hline \end{array}$$

21
$$\begin{array}{r} 2\ 1 \\ \div\quad 3 \\ \hline \end{array}$$

22
$$\begin{array}{r} 3 \\ \div\quad 3 \\ \hline \end{array}$$

23
$$\begin{array}{r} 4\ 0 \\ \div\quad 4 \\ \hline \end{array}$$

24
$$\begin{array}{r} 2\ 0 \\ \div\quad 5 \\ \hline \end{array}$$

25
$$\begin{array}{r} 1\ 2 \\ \div\quad 4 \\ \hline \end{array}$$

26
$$\begin{array}{r} 1\ 0 \\ \div\quad 5 \\ \hline \end{array}$$

27
$$\begin{array}{r} 2 \\ \div\quad 2 \\ \hline \end{array}$$

28
$$\begin{array}{r} 6 \\ \div\quad 2 \\ \hline \end{array}$$

29
$$\begin{array}{r} 2\ 4 \\ \div\quad 4 \\ \hline \end{array}$$

30
$$\begin{array}{r} 3\ 0 \\ \div\quad 5 \\ \hline \end{array}$$

31
$$\begin{array}{r} 4 \\ \div\quad 4 \\ \hline \end{array}$$

32
$$\begin{array}{r} 6 \\ \div\quad 3 \\ \hline \end{array}$$

33
$$\begin{array}{r} 9 \\ \div\quad 3 \\ \hline \end{array}$$

34
$$\begin{array}{r} 1\ 4 \\ \div\quad 2 \\ \hline \end{array}$$

35
$$\begin{array}{r} 4 \\ \div\quad 2 \\ \hline \end{array}$$

36
$$\begin{array}{r} 1\ 2 \\ \div\quad 2 \\ \hline \end{array}$$

SET I Date: _____ Start: _____ Finish: _____ Score: _____

1)
$$4 \div 2$$

2)
$$8 \div 4$$

3)
$$36 \div 4$$

4)
$$32 \div 4$$

5)
$$16 \div 4$$

6)
$$12 \div 3$$

7)
$$24 \div 3$$

8)
$$15 \div 3$$

9)
$$50 \div 5$$

10)
$$5 \div 5$$

11)
$$10 \div 5$$

12)
$$2 \div 2$$

13)
$$27 \div 3$$

14)
$$25 \div 5$$

15)
$$18 \div 3$$

16)
$$24 \div 4$$

17)
$$16 \div 2$$

18)
$$20 \div 2$$

SET II Date: _____ Start: _____ Finish: _____ Score: _____

19)
$$8 \div 2$$

20)
$$12 \div 4$$

21)
$$10 \div 2$$

22)
$$20 \div 4$$

23)
$$28 \div 4$$

24)
$$35 \div 5$$

25)
$$14 \div 2$$

26)
$$12 \div 2$$

27)
$$9 \div 3$$

28)
$$30 \div 5$$

29)
$$30 \div 3$$

30)
$$15 \div 5$$

31)
$$20 \div 5$$

32)
$$45 \div 5$$

33)
$$4 \div 4$$

34)
$$6 \div 2$$

35)
$$40 \div 5$$

36)
$$18 \div 2$$

SET I   Date: _____   Start: _____   Finish: _____   Score: _____

1	2	3	4	5	6
2 4 ÷ 3	8 ÷ 4	1 2 ÷ 4	4 ÷ 2	1 5 ÷ 5	1 0 ÷ 5

7	8	9	10	11	12
6 ÷ 2	8 ÷ 2	4 ÷ 4	2 0 ÷ 5	4 0 ÷ 4	4 5 ÷ 5

13	14	15	16	17	18
2 4 ÷ 4	2 1 ÷ 3	3 0 ÷ 3	1 2 ÷ 3	9 ÷ 3	1 6 ÷ 2

SET II   Date: _____   Start: _____   Finish: _____   Score: _____

19	20	21	22	23	24
3 2 ÷ 4	3 0 ÷ 5	5 0 ÷ 5	1 6 ÷ 4	3 5 ÷ 5	6 ÷ 3

25	26	27	28	29	30
2 7 ÷ 3	4 0 ÷ 5	2 ÷ 2	1 8 ÷ 3	5 ÷ 5	1 8 ÷ 2

31	32	33	34	35	36
1 4 ÷ 2	3 6 ÷ 4	2 0 ÷ 4	1 2 ÷ 2	2 5 ÷ 5	2 0 ÷ 2

Division Facts

SET I Date: _____ Start: _____ Finish: _____ Score: _____

1	2	3	4	5	6
3 0 ÷ 5	1 8 ÷ 3	2 7 ÷ 3	2 4 ÷ 4	2 0 ÷ 2	5 0 ÷ 5

7	8	9	10	11	12
2 1 ÷ 3	1 5 ÷ 3	2 4 ÷ 3	4 ÷ 2	1 2 ÷ 3	1 6 ÷ 4

13	14	15	16	17	18
3 5 ÷ 5	3 ÷ 3	4 5 ÷ 5	1 2 ÷ 4	2 0 ÷ 4	1 5 ÷ 5

SET II Date: _____ Start: _____ Finish: _____ Score: _____

19	20	21	22	23	24
3 6 ÷ 4	2 5 ÷ 5	4 0 ÷ 5	1 0 ÷ 2	5 ÷ 5	9 ÷ 3

25	26	27	28	29	30
6 ÷ 2	2 ÷ 2	8 ÷ 4	6 ÷ 3	1 0 ÷ 5	3 2 ÷ 4

31	32	33	34	35	36
1 8 ÷ 2	1 6 ÷ 2	2 8 ÷ 4	1 4 ÷ 2	8 ÷ 2	3 0 ÷ 3

SET I Date: _____ Start: _____ Finish: _____ Score: _____

(1) $$20 \div 2$$	(2) $$32 \div 4$$	(3) $$18 \div 2$$
(4) $$20 \div 5$$	(5) $$24 \div 4$$	(6) $$36 \div 4$$

(7) $$12 \div 3$$	(8) $$2 \div 2$$	(9) $$16 \div 2$$
(10) $$5 \div 5$$	(11) $$4 \div 2$$	(12) $$25 \div 5$$

(13) $$35 \div 5$$	(14) $$21 \div 3$$	(15) $$30 \div 5$$
(16) $$12 \div 4$$	(17) $$24 \div 3$$	(18) $$40 \div 5$$

SET II Date: _____ Start: _____ Finish: _____ Score: _____

(19) $$50 \div 5$$	(20) $$15 \div 3$$	(21) $$18 \div 3$$
(22) $$12 \div 2$$	(23) $$30 \div 3$$	(24) $$8 \div 4$$

(25) $$16 \div 4$$	(26) $$8 \div 2$$	(27) $$45 \div 5$$
(28) $$10 \div 5$$	(29) $$20 \div 4$$	(30) $$14 \div 2$$

(31) $$6 \div 2$$	(32) $$9 \div 3$$	(33) $$15 \div 5$$
(34) $$10 \div 2$$	(35) $$27 \div 3$$	(36) $$28 \div 4$$

SET I Date: _____ Start: _____ Finish: _____ Score: _____

1	**2**	**3**	**4**	**5**	**6**
4 0 ÷ 4	3 6 ÷ 4	1 5 ÷ 3	2 7 ÷ 3	2 4 ÷ 3	8 ÷ 2
7	**8**	**9**	**10**	**11**	**12**
9 ÷ 3	5 0 ÷ 5	1 2 ÷ 3	5 ÷ 5	2 ÷ 2	1 8 ÷ 2
13	**14**	**15**	**16**	**17**	**18**
2 1 ÷ 3	1 2 ÷ 4	2 5 ÷ 5	8 ÷ 4	4 0 ÷ 5	1 2 ÷ 2

SET II Date: _____ Start: _____ Finish: _____ Score: _____

19	**20**	**21**	**22**	**23**	**24**
2 0 ÷ 2	1 6 ÷ 4	2 0 ÷ 5	3 ÷ 3	4 5 ÷ 5	2 8 ÷ 4
25	**26**	**27**	**28**	**29**	**30**
1 0 ÷ 2	1 4 ÷ 2	3 5 ÷ 5	6 ÷ 3	3 0 ÷ 3	1 0 ÷ 5
31	**32**	**33**	**34**	**35**	**36**
3 2 ÷ 4	6 ÷ 2	3 0 ÷ 5	1 6 ÷ 2	1 5 ÷ 5	2 4 ÷ 4

SET I   Date: _____   Start: _____   Finish: _____   Score: _____

1
$$6 \div 6$$

2
$$3\ 6 \div 6$$

3
$$1\ 2 \div 6$$

4
$$3\ 0 \div 6$$

5
$$6\ 0 \div 6$$

6
$$4\ 2 \div 6$$

7
$$1\ 8 \div 6$$

8
$$2\ 4 \div 6$$

9
$$5\ 4 \div 6$$

10
$$4\ 8 \div 6$$

11
$$3\ 6 \div 6$$

12
$$5\ 4 \div 6$$

13
$$4\ 8 \div 6$$

14
$$3\ 0 \div 6$$

15
$$1\ 8 \div 6$$

16
$$4\ 2 \div 6$$

17
$$6\ 0 \div 6$$

18
$$6 \div 6$$

SET II   Date: _____   Start: _____   Finish: _____   Score: _____

19
$$2\ 4 \div 6$$

20
$$1\ 2 \div 6$$

21
$$6\ 0 \div 6$$

22
$$1\ 2 \div 6$$

23
$$6 \div 6$$

24
$$5\ 4 \div 6$$

25
$$4\ 2 \div 6$$

26
$$2\ 4 \div 6$$

27
$$4\ 8 \div 6$$

28
$$3\ 0 \div 6$$

29
$$1\ 8 \div 6$$

30
$$3\ 6 \div 6$$

31
$$1\ 8 \div 6$$

32
$$2\ 4 \div 6$$

33
$$4\ 2 \div 6$$

34
$$1\ 2 \div 6$$

35
$$3\ 6 \div 6$$

36
$$5\ 4 \div 6$$

SET I   Date: _____   Start: _____   Finish: _____   Score: _____

1
$$\begin{array}{r} 2\ 4 \\ \div\quad 6 \\ \hline \end{array}$$

2
$$\begin{array}{r} 4\ 2 \\ \div\quad 6 \\ \hline \end{array}$$

3
$$\begin{array}{r} 1\ 2 \\ \div\quad 6 \\ \hline \end{array}$$

4
$$\begin{array}{r} 6\ 0 \\ \div\quad 6 \\ \hline \end{array}$$

5
$$\begin{array}{r} 3\ 0 \\ \div\quad 6 \\ \hline \end{array}$$

6
$$\begin{array}{r} 4\ 8 \\ \div\quad 6 \\ \hline \end{array}$$

7
$$\begin{array}{r} 3\ 6 \\ \div\quad 6 \\ \hline \end{array}$$

8
$$\begin{array}{r} 6 \\ \div\quad 6 \\ \hline \end{array}$$

9
$$\begin{array}{r} 5\ 4 \\ \div\quad 6 \\ \hline \end{array}$$

10
$$\begin{array}{r} 1\ 8 \\ \div\quad 6 \\ \hline \end{array}$$

11
$$\begin{array}{r} 2\ 4 \\ \div\quad 6 \\ \hline \end{array}$$

12
$$\begin{array}{r} 1\ 2 \\ \div\quad 6 \\ \hline \end{array}$$

13
$$\begin{array}{r} 4\ 2 \\ \div\quad 6 \\ \hline \end{array}$$

14
$$\begin{array}{r} 3\ 6 \\ \div\quad 6 \\ \hline \end{array}$$

15
$$\begin{array}{r} 4\ 8 \\ \div\quad 6 \\ \hline \end{array}$$

16
$$\begin{array}{r} 5\ 4 \\ \div\quad 6 \\ \hline \end{array}$$

17
$$\begin{array}{r} 1\ 8 \\ \div\quad 6 \\ \hline \end{array}$$

18
$$\begin{array}{r} 6\ 0 \\ \div\quad 6 \\ \hline \end{array}$$

SET II   Date: _____   Start: _____   Finish: _____   Score: _____

19
$$\begin{array}{r} 3\ 0 \\ \div\quad 6 \\ \hline \end{array}$$

20
$$\begin{array}{r} 6 \\ \div\quad 6 \\ \hline \end{array}$$

21
$$\begin{array}{r} 2\ 4 \\ \div\quad 6 \\ \hline \end{array}$$

22
$$\begin{array}{r} 1\ 8 \\ \div\quad 6 \\ \hline \end{array}$$

23
$$\begin{array}{r} 6\ 0 \\ \div\quad 6 \\ \hline \end{array}$$

24
$$\begin{array}{r} 6 \\ \div\quad 6 \\ \hline \end{array}$$

25
$$\begin{array}{r} 4\ 8 \\ \div\quad 6 \\ \hline \end{array}$$

26
$$\begin{array}{r} 5\ 4 \\ \div\quad 6 \\ \hline \end{array}$$

27
$$\begin{array}{r} 3\ 0 \\ \div\quad 6 \\ \hline \end{array}$$

28
$$\begin{array}{r} 1\ 2 \\ \div\quad 6 \\ \hline \end{array}$$

29
$$\begin{array}{r} 3\ 6 \\ \div\quad 6 \\ \hline \end{array}$$

30
$$\begin{array}{r} 4\ 2 \\ \div\quad 6 \\ \hline \end{array}$$

31
$$\begin{array}{r} 3\ 6 \\ \div\quad 6 \\ \hline \end{array}$$

32
$$\begin{array}{r} 6 \\ \div\quad 6 \\ \hline \end{array}$$

33
$$\begin{array}{r} 5\ 4 \\ \div\quad 6 \\ \hline \end{array}$$

34
$$\begin{array}{r} 3\ 0 \\ \div\quad 6 \\ \hline \end{array}$$

35
$$\begin{array}{r} 4\ 2 \\ \div\quad 6 \\ \hline \end{array}$$

36
$$\begin{array}{r} 6\ 0 \\ \div\quad 6 \\ \hline \end{array}$$

SET I   Date: _____   Start: _____   Finish: _____   Score: _____

1
$$2\ 4$$
$$\div\ \ 6$$

2
$$6$$
$$\div\ \ 6$$

3
$$3\ 6$$
$$\div\ \ 6$$

4
$$4\ 8$$
$$\div\ \ 6$$

5
$$6\ 0$$
$$\div\ \ 6$$

6
$$4\ 2$$
$$\div\ \ 6$$

7
$$3\ 0$$
$$\div\ \ 6$$

8
$$5\ 4$$
$$\div\ \ 6$$

9
$$1\ 8$$
$$\div\ \ 6$$

10
$$1\ 2$$
$$\div\ \ 6$$

11
$$4\ 8$$
$$\div\ \ 6$$

12
$$3\ 6$$
$$\div\ \ 6$$

13
$$2\ 4$$
$$\div\ \ 6$$

14
$$6\ 0$$
$$\div\ \ 6$$

15
$$4\ 2$$
$$\div\ \ 6$$

16
$$1\ 2$$
$$\div\ \ 6$$

17
$$1\ 8$$
$$\div\ \ 6$$

18
$$6$$
$$\div\ \ 6$$

SET II   Date: _____   Start: _____   Finish: _____   Score: _____

19
$$3\ 0$$
$$\div\ \ 6$$

20
$$5\ 4$$
$$\div\ \ 6$$

21
$$1\ 2$$
$$\div\ \ 6$$

22
$$4\ 2$$
$$\div\ \ 6$$

23
$$2\ 4$$
$$\div\ \ 6$$

24
$$3\ 0$$
$$\div\ \ 6$$

25
$$5\ 4$$
$$\div\ \ 6$$

26
$$3\ 6$$
$$\div\ \ 6$$

27
$$4\ 8$$
$$\div\ \ 6$$

28
$$6$$
$$\div\ \ 6$$

29
$$1\ 8$$
$$\div\ \ 6$$

30
$$6\ 0$$
$$\div\ \ 6$$

31
$$3\ 6$$
$$\div\ \ 6$$

32
$$3\ 0$$
$$\div\ \ 6$$

33
$$2\ 4$$
$$\div\ \ 6$$

34
$$6$$
$$\div\ \ 6$$

35
$$5\ 4$$
$$\div\ \ 6$$

36
$$4\ 8$$
$$\div\ \ 6$$

SET I Date: _____ Start: _____ Finish: _____ Score: _____

1.
$$6 \div 6$$

2.
$$12 \div 6$$

3.
$$30 \div 6$$

4.
$$24 \div 6$$

5.
$$18 \div 6$$

6.
$$48 \div 6$$

7.
$$60 \div 6$$

8.
$$54 \div 6$$

9.
$$36 \div 6$$

10.
$$42 \div 6$$

11.
$$36 \div 6$$

12.
$$42 \div 6$$

13.
$$30 \div 6$$

14.
$$24 \div 6$$

15.
$$12 \div 6$$

16.
$$6 \div 6$$

17.
$$60 \div 6$$

18.
$$48 \div 6$$

SET II Date: _____ Start: _____ Finish: _____ Score: _____

19.
$$54 \div 6$$

20.
$$18 \div 6$$

21.
$$42 \div 6$$

22.
$$24 \div 6$$

23.
$$6 \div 6$$

24.
$$48 \div 6$$

25.
$$60 \div 6$$

26.
$$30 \div 6$$

27.
$$12 \div 6$$

28.
$$18 \div 6$$

29.
$$36 \div 6$$

30.
$$54 \div 6$$

31.
$$12 \div 6$$

32.
$$30 \div 6$$

33.
$$6 \div 6$$

34.
$$42 \div 6$$

35.
$$48 \div 6$$

36.
$$18 \div 6$$

Division Facts

SET I Date: _____ Start: _____ Finish: _____ Score: _____

1
4 2
÷   7

2
7
÷   7

3
7 0
÷   7

4
2 8
÷   7

5
2 1
÷   7

6
1 4
÷   7

7
6 3
÷   7

8
5 6
÷   7

9
3 5
÷   7

10
4 9
÷   7

11
6 3
÷   7

12
2 1
÷   7

13
2 8
÷   7

14
5 6
÷   7

15
7
÷   7

16
7 0
÷   7

17
4 9
÷   7

18
3 5
÷   7

SET II Date: _____ Start: _____ Finish: _____ Score: _____

19
4 2
÷   7

20
1 4
÷   7

21
3 5
÷   7

22
4 9
÷   7

23
2 8
÷   7

24
5 6
÷   7

25
1 4
÷   7

26
7
÷   7

27
7 0
÷   7

28
2 1
÷   7

29
4 2
÷   7

30
6 3
÷   7

31
4 2
÷   7

32
7
÷   7

33
7 0
÷   7

34
5 6
÷   7

35
4 9
÷   7

36
1 4
÷   7

SET I   Date: _____   Start: _____   Finish: _____   Score: _____

1	2	3	4	5	6
7 0 ÷ 7	6 3 ÷ 7	1 4 ÷ 7	3 5 ÷ 7	5 6 ÷ 7	4 2 ÷ 7

7	8	9	10	11	12
2 8 ÷ 7	4 9 ÷ 7	2 1 ÷ 7	7 ÷ 7	7 ÷ 7	5 6 ÷ 7

13	14	15	16	17	18
1 4 ÷ 7	2 8 ÷ 7	6 3 ÷ 7	3 5 ÷ 7	7 0 ÷ 7	4 9 ÷ 7

SET II   Date: _____   Start: _____   Finish: _____   Score: _____

19	20	21	22	23	24
2 1 ÷ 7	4 2 ÷ 7	7 0 ÷ 7	7 ÷ 7	2 1 ÷ 7	1 4 ÷ 7

25	26	27	28	29	30
4 9 ÷ 7	3 5 ÷ 7	6 3 ÷ 7	5 6 ÷ 7	2 8 ÷ 7	4 2 ÷ 7

31	32	33	34	35	36
3 5 ÷ 7	1 4 ÷ 7	6 3 ÷ 7	2 1 ÷ 7	5 6 ÷ 7	7 0 ÷ 7

Division Facts

SET I Date: _____ Start: _____ Finish: _____ Score: _____

1	2	3	4	5	6
2 1 ÷ 7	4 2 ÷ 7	3 5 ÷ 7	1 4 ÷ 7	4 9 ÷ 7	6 3 ÷ 7

7	8	9	10	11	12
2 8 ÷ 7	7 0 ÷ 7	5 6 ÷ 7	7 ÷ 7	7 0 ÷ 7	2 8 ÷ 7

13	14	15	16	17	18
7 ÷ 7	2 1 ÷ 7	6 3 ÷ 7	4 2 ÷ 7	1 4 ÷ 7	4 9 ÷ 7

SET II Date: _____ Start: _____ Finish: _____ Score: _____

19	20	21	22	23	24
3 5 ÷ 7	5 6 ÷ 7	7 ÷ 7	3 5 ÷ 7	5 6 ÷ 7	2 8 ÷ 7

25	26	27	28	29	30
4 9 ÷ 7	2 1 ÷ 7	4 2 ÷ 7	7 0 ÷ 7	1 4 ÷ 7	6 3 ÷ 7

31	32	33	34	35	36
7 0 ÷ 7	2 8 ÷ 7	4 9 ÷ 7	1 4 ÷ 7	7 ÷ 7	3 5 ÷ 7

Division Facts

SET I   Date: _____   Start: _____   Finish: _____   Score: _____

1.
$$7 \div 7$$

2.
$$49 \div 7$$

3.
$$56 \div 7$$

4.
$$63 \div 7$$

5.
$$42 \div 7$$

6.
$$14 \div 7$$

7.
$$35 \div 7$$

8.
$$21 \div 7$$

9.
$$70 \div 7$$

10.
$$28 \div 7$$

11.
$$14 \div 7$$

12.
$$49 \div 7$$

13.
$$70 \div 7$$

14.
$$28 \div 7$$

15.
$$7 \div 7$$

16.
$$21 \div 7$$

17.
$$35 \div 7$$

18.
$$56 \div 7$$

SET II   Date: _____   Start: _____   Finish: _____   Score: _____

19.
$$63 \div 7$$

20.
$$42 \div 7$$

21.
$$56 \div 7$$

22.
$$7 \div 7$$

23.
$$42 \div 7$$

24.
$$28 \div 7$$

25.
$$21 \div 7$$

26.
$$14 \div 7$$

27.
$$35 \div 7$$

28.
$$49 \div 7$$

29.
$$70 \div 7$$

30.
$$63 \div 7$$

31.
$$28 \div 7$$

32.
$$56 \div 7$$

33.
$$42 \div 7$$

34.
$$63 \div 7$$

35.
$$35 \div 7$$

36.
$$49 \div 7$$

SET I Date: _____ Start: _____ Finish: _____ Score: _____

1. 2 8 ÷ 7
2. 6 ÷ 6
3. 6 0 ÷ 6
4. 1 4 ÷ 7
5. 6 3 ÷ 7
6. 4 9 ÷ 7

7. 2 4 ÷ 6
8. 7 0 ÷ 7
9. 1 2 ÷ 6
10. 5 6 ÷ 7
11. 4 2 ÷ 7
12. 1 8 ÷ 6

13. 3 5 ÷ 7
14. 3 0 ÷ 6
15. 7 ÷ 7
16. 5 4 ÷ 6
17. 4 2 ÷ 6
18. 3 6 ÷ 6

SET II Date: _____ Start: _____ Finish: _____ Score: _____

19. 4 8 ÷ 6
20. 2 1 ÷ 7
21. 2 8 ÷ 7
22. 6 ÷ 6
23. 6 0 ÷ 6
24. 1 4 ÷ 7

25. 6 3 ÷ 7
26. 4 9 ÷ 7
27. 2 4 ÷ 6
28. 7 0 ÷ 7
29. 1 2 ÷ 6
30. 5 6 ÷ 7

31. 4 2 ÷ 7
32. 1 8 ÷ 6
33. 3 5 ÷ 7
34. 3 0 ÷ 6
35. 7 ÷ 7
36. 5 4 ÷ 6

Review: Dividing by 6 and 7

SET I   Date: _____   Start: _____   Finish: _____   Score: _____

1. 6 ÷ 6
2. 4 2 ÷ 6
3. 5 6 ÷ 7
4. 7 ÷ 7
5. 2 4 ÷ 6
6. 1 4 ÷ 7
7. 4 8 ÷ 6
8. 2 1 ÷ 7
9. 6 3 ÷ 7
10. 1 8 ÷ 6
11. 1 2 ÷ 6
12. 4 2 ÷ 7
13. 3 6 ÷ 6
14. 4 9 ÷ 7
15. 3 5 ÷ 7
16. 2 8 ÷ 7
17. 5 4 ÷ 6
18. 3 0 ÷ 6

SET II   Date: _____   Start: _____   Finish: _____   Score: _____

19. 6 0 ÷ 6
20. 7 0 ÷ 7
21. 6 ÷ 6
22. 4 2 ÷ 6
23. 5 6 ÷ 7
24. 7 ÷ 7
25. 2 4 ÷ 6
26. 1 4 ÷ 7
27. 4 8 ÷ 6
28. 2 1 ÷ 7
29. 6 3 ÷ 7
30. 1 8 ÷ 6
31. 1 2 ÷ 6
32. 4 2 ÷ 7
33. 3 6 ÷ 6
34. 4 9 ÷ 7
35. 3 5 ÷ 7
36. 2 8 ÷ 7

50   Division Facts

SET I

Date: _____ Start: _____ Finish: _____ Score: _____

1
$$\begin{array}{r} 1\ 2 \\ \div\quad 6 \\ \hline \end{array}$$

2
$$\begin{array}{r} 3\ 0 \\ \div\quad 6 \\ \hline \end{array}$$

3
$$\begin{array}{r} 4\ 2 \\ \div\quad 6 \\ \hline \end{array}$$

4
$$\begin{array}{r} 4\ 2 \\ \div\quad 7 \\ \hline \end{array}$$

5
$$\begin{array}{r} 3\ 6 \\ \div\quad 6 \\ \hline \end{array}$$

6
$$\begin{array}{r} 2\ 4 \\ \div\quad 6 \\ \hline \end{array}$$

7
$$\begin{array}{r} 7 \\ \div\quad 7 \\ \hline \end{array}$$

8
$$\begin{array}{r} 5\ 4 \\ \div\quad 6 \\ \hline \end{array}$$

9
$$\begin{array}{r} 3\ 5 \\ \div\quad 7 \\ \hline \end{array}$$

10
$$\begin{array}{r} 4\ 8 \\ \div\quad 6 \\ \hline \end{array}$$

11
$$\begin{array}{r} 7\ 0 \\ \div\quad 7 \\ \hline \end{array}$$

12
$$\begin{array}{r} 2\ 8 \\ \div\quad 7 \\ \hline \end{array}$$

13
$$\begin{array}{r} 1\ 8 \\ \div\quad 6 \\ \hline \end{array}$$

14
$$\begin{array}{r} 1\ 4 \\ \div\quad 7 \\ \hline \end{array}$$

15
$$\begin{array}{r} 6\ 0 \\ \div\quad 6 \\ \hline \end{array}$$

16
$$\begin{array}{r} 5\ 6 \\ \div\quad 7 \\ \hline \end{array}$$

17
$$\begin{array}{r} 6 \\ \div\quad 6 \\ \hline \end{array}$$

18
$$\begin{array}{r} 2\ 1 \\ \div\quad 7 \\ \hline \end{array}$$

SET II

Date: _____ Start: _____ Finish: _____ Score: _____

19
$$\begin{array}{r} 6\ 3 \\ \div\quad 7 \\ \hline \end{array}$$

20
$$\begin{array}{r} 4\ 9 \\ \div\quad 7 \\ \hline \end{array}$$

21
$$\begin{array}{r} 1\ 2 \\ \div\quad 6 \\ \hline \end{array}$$

22
$$\begin{array}{r} 3\ 0 \\ \div\quad 6 \\ \hline \end{array}$$

23
$$\begin{array}{r} 4\ 2 \\ \div\quad 6 \\ \hline \end{array}$$

24
$$\begin{array}{r} 4\ 2 \\ \div\quad 7 \\ \hline \end{array}$$

25
$$\begin{array}{r} 3\ 6 \\ \div\quad 6 \\ \hline \end{array}$$

26
$$\begin{array}{r} 2\ 4 \\ \div\quad 6 \\ \hline \end{array}$$

27
$$\begin{array}{r} 7 \\ \div\quad 7 \\ \hline \end{array}$$

28
$$\begin{array}{r} 5\ 4 \\ \div\quad 6 \\ \hline \end{array}$$

29
$$\begin{array}{r} 3\ 5 \\ \div\quad 7 \\ \hline \end{array}$$

30
$$\begin{array}{r} 4\ 8 \\ \div\quad 6 \\ \hline \end{array}$$

31
$$\begin{array}{r} 7\ 0 \\ \div\quad 7 \\ \hline \end{array}$$

32
$$\begin{array}{r} 2\ 8 \\ \div\quad 7 \\ \hline \end{array}$$

33
$$\begin{array}{r} 1\ 8 \\ \div\quad 6 \\ \hline \end{array}$$

34
$$\begin{array}{r} 1\ 4 \\ \div\quad 7 \\ \hline \end{array}$$

35
$$\begin{array}{r} 6\ 0 \\ \div\quad 6 \\ \hline \end{array}$$

36
$$\begin{array}{r} 5\ 6 \\ \div\quad 7 \\ \hline \end{array}$$

SET I Date: _____ Start: _____ Finish: _____ Score: _____

1	2	3	4	5	6
1 2 ÷ 6	5 4 ÷ 6	6 3 ÷ 7	3 0 ÷ 6	4 2 ÷ 6	2 1 ÷ 7

7	8	9	10	11	12
6 0 ÷ 6	1 8 ÷ 6	2 8 ÷ 7	2 4 ÷ 6	4 2 ÷ 7	1 4 ÷ 7

13	14	15	16	17	18
5 6 ÷ 7	7 0 ÷ 7	3 5 ÷ 7	7 ÷ 7	4 8 ÷ 6	6 ÷ 6

SET II Date: _____ Start: _____ Finish: _____ Score: _____

19	20	21	22	23	24
3 6 ÷ 6	4 9 ÷ 7	1 2 ÷ 6	5 4 ÷ 6	6 3 ÷ 7	3 0 ÷ 6

25	26	27	28	29	30
4 2 ÷ 6	2 1 ÷ 7	6 0 ÷ 6	1 8 ÷ 6	2 8 ÷ 7	2 4 ÷ 6

31	32	33	34	35	36
4 2 ÷ 7	1 4 ÷ 7	5 6 ÷ 7	7 0 ÷ 7	3 5 ÷ 7	7 ÷ 7

Division Facts

SET I   Date: _____   Start: _____   Finish: _____   Score: _____

1	2	3	4	5	6
6 3 ÷ 7	3 5 ÷ 7	3 6 ÷ 6	1 4 ÷ 7	7 ÷ 7	5 6 ÷ 7

7	8	9	10	11	12
6 ÷ 6	6 0 ÷ 6	4 9 ÷ 7	4 8 ÷ 6	5 4 ÷ 6	4 2 ÷ 6

13	14	15	16	17	18
7 0 ÷ 7	2 8 ÷ 7	1 2 ÷ 6	3 0 ÷ 6	4 2 ÷ 7	1 8 ÷ 6

SET II   Date: _____   Start: _____   Finish: _____   Score: _____

19	20	21	22	23	24
2 1 ÷ 7	2 4 ÷ 6	6 3 ÷ 7	3 5 ÷ 7	3 6 ÷ 6	1 4 ÷ 7

25	26	27	28	29	30
7 ÷ 7	5 6 ÷ 7	6 ÷ 6	6 0 ÷ 6	4 9 ÷ 7	4 8 ÷ 6

31	32	33	34	35	36
5 4 ÷ 6	4 2 ÷ 6	7 0 ÷ 7	2 8 ÷ 7	1 2 ÷ 6	3 0 ÷ 6

SET I Date: _____ Start: _____ Finish: _____ Score: _____

1.
$$3 \ 6 \div 6$$

2.
$$3 \ 5 \div 7$$

3.
$$6 \ 0 \div 6$$

4.
$$4 \ 2 \div 6$$

5.
$$2 \ 1 \div 7$$

6.
$$2 \ 4 \div 6$$

7.
$$7 \div 7$$

8.
$$6 \div 6$$

9.
$$5 \ 4 \div 6$$

10.
$$4 \ 8 \div 6$$

11.
$$1 \ 8 \div 6$$

12.
$$2 \ 8 \div 7$$

13.
$$4 \ 2 \div 7$$

14.
$$6 \ 3 \div 7$$

15.
$$3 \ 0 \div 6$$

16.
$$7 \ 0 \div 7$$

17.
$$1 \ 2 \div 6$$

18.
$$5 \ 6 \div 7$$

SET II Date: _____ Start: _____ Finish: _____ Score: _____

19.
$$1 \ 4 \div 7$$

20.
$$4 \ 9 \div 7$$

21.
$$3 \ 6 \div 6$$

22.
$$3 \ 5 \div 7$$

23.
$$6 \ 0 \div 6$$

24.
$$4 \ 2 \div 6$$

25.
$$2 \ 1 \div 7$$

26.
$$2 \ 4 \div 6$$

27.
$$7 \div 7$$

28.
$$6 \div 6$$

29.
$$5 \ 4 \div 6$$

30.
$$4 \ 8 \div 6$$

31.
$$1 \ 8 \div 6$$

32.
$$2 \ 8 \div 7$$

33.
$$4 \ 2 \div 7$$

34.
$$6 \ 3 \div 7$$

35.
$$3 \ 0 \div 6$$

36.
$$7 \ 0 \div 7$$

Division Facts

SET I   Date: _____   Start: _____   Finish: _____   Score: _____

1
$$4\ 8$$
$$\div\quad 8$$

2
$$1\ 6$$
$$\div\quad 8$$

3
$$3\ 2$$
$$\div\quad 8$$

4
$$8\ 0$$
$$\div\quad 8$$

5
$$6\ 4$$
$$\div\quad 8$$

6
$$5\ 6$$
$$\div\quad 8$$

7
$$4\ 0$$
$$\div\quad 8$$

8
$$7\ 2$$
$$\div\quad 8$$

9
$$2\ 4$$
$$\div\quad 8$$

10
$$8$$
$$\div\quad 8$$

11
$$1\ 6$$
$$\div\quad 8$$

12
$$6\ 4$$
$$\div\quad 8$$

13
$$7\ 2$$
$$\div\quad 8$$

14
$$4\ 8$$
$$\div\quad 8$$

15
$$8$$
$$\div\quad 8$$

16
$$5\ 6$$
$$\div\quad 8$$

17
$$3\ 2$$
$$\div\quad 8$$

18
$$4\ 0$$
$$\div\quad 8$$

SET II   Date: _____   Start: _____   Finish: _____   Score: _____

19
$$2\ 4$$
$$\div\quad 8$$

20
$$8\ 0$$
$$\div\quad 8$$

21
$$7\ 2$$
$$\div\quad 8$$

22
$$3\ 2$$
$$\div\quad 8$$

23
$$5\ 6$$
$$\div\quad 8$$

24
$$8$$
$$\div\quad 8$$

25
$$2\ 4$$
$$\div\quad 8$$

26
$$1\ 6$$
$$\div\quad 8$$

27
$$8\ 0$$
$$\div\quad 8$$

28
$$4\ 8$$
$$\div\quad 8$$

29
$$4\ 0$$
$$\div\quad 8$$

30
$$6\ 4$$
$$\div\quad 8$$

31
$$8\ 0$$
$$\div\quad 8$$

32
$$8$$
$$\div\quad 8$$

33
$$1\ 6$$
$$\div\quad 8$$

34
$$4\ 8$$
$$\div\quad 8$$

35
$$7\ 2$$
$$\div\quad 8$$

36
$$4\ 0$$
$$\div\quad 8$$

Division Facts

SET I   Date: _____   Start: _____   Finish: _____   Score: _____

1	2	3	4	5	6
2 4 ÷ 8	5 6 ÷ 8	7 2 ÷ 8	8 0 ÷ 8	4 8 ÷ 8	8 ÷ 8

7	8	9	10	11	12
4 0 ÷ 8	6 4 ÷ 8	1 6 ÷ 8	3 2 ÷ 8	5 6 ÷ 8	4 8 ÷ 8

13	14	15	16	17	18
6 4 ÷ 8	4 0 ÷ 8	8 0 ÷ 8	1 6 ÷ 8	2 4 ÷ 8	8 ÷ 8

SET II   Date: _____   Start: _____   Finish: _____   Score: _____

19	20	21	22	23	24
3 2 ÷ 8	7 2 ÷ 8	5 6 ÷ 8	1 6 ÷ 8	3 2 ÷ 8	6 4 ÷ 8

25	26	27	28	29	30
4 0 ÷ 8	2 4 ÷ 8	8 0 ÷ 8	4 8 ÷ 8	8 ÷ 8	7 2 ÷ 8

31	32	33	34	35	36
8 0 ÷ 8	2 4 ÷ 8	4 8 ÷ 8	3 2 ÷ 8	8 ÷ 8	1 6 ÷ 8

SET I   Date: _____   Start: _____   Finish: _____   Score: _____

1
$$8\ 0 \div 8$$

2
$$3\ 2 \div 8$$

3
$$6\ 4 \div 8$$

4
$$4\ 0 \div 8$$

5
$$5\ 6 \div 8$$

6
$$1\ 6 \div 8$$

7
$$2\ 4 \div 8$$

8
$$7\ 2 \div 8$$

9
$$4\ 8 \div 8$$

10
$$8 \div 8$$

11
$$7\ 2 \div 8$$

12
$$4\ 8 \div 8$$

13
$$4\ 0 \div 8$$

14
$$8 \div 8$$

15
$$2\ 4 \div 8$$

16
$$8\ 0 \div 8$$

17
$$3\ 2 \div 8$$

18
$$5\ 6 \div 8$$

SET II   Date: _____   Start: _____   Finish: _____   Score: _____

19
$$1\ 6 \div 8$$

20
$$6\ 4 \div 8$$

21
$$4\ 8 \div 8$$

22
$$1\ 6 \div 8$$

23
$$2\ 4 \div 8$$

24
$$6\ 4 \div 8$$

25
$$3\ 2 \div 8$$

26
$$4\ 0 \div 8$$

27
$$8 \div 8$$

28
$$7\ 2 \div 8$$

29
$$8\ 0 \div 8$$

30
$$5\ 6 \div 8$$

31
$$2\ 4 \div 8$$

32
$$8\ 0 \div 8$$

33
$$8 \div 8$$

34
$$6\ 4 \div 8$$

35
$$1\ 6 \div 8$$

36
$$4\ 8 \div 8$$

SET I   Date: _____   Start: _____   Finish: _____   Score: _____

1)
$$32 \div 8$$

2)
$$80 \div 8$$

3)
$$24 \div 8$$

4)
$$16 \div 8$$

5)
$$56 \div 8$$

6)
$$40 \div 8$$

7)
$$72 \div 8$$

8)
$$64 \div 8$$

9)
$$48 \div 8$$

10)
$$8 \div 8$$

11)
$$24 \div 8$$

12)
$$40 \div 8$$

13)
$$72 \div 8$$

14)
$$80 \div 8$$

15)
$$32 \div 8$$

16)
$$48 \div 8$$

17)
$$64 \div 8$$

18)
$$8 \div 8$$

SET II   Date: _____   Start: _____   Finish: _____   Score: _____

19)
$$16 \div 8$$

20)
$$56 \div 8$$

21)
$$8 \div 8$$

22)
$$16 \div 8$$

23)
$$56 \div 8$$

24)
$$80 \div 8$$

25)
$$48 \div 8$$

26)
$$24 \div 8$$

27)
$$64 \div 8$$

28)
$$72 \div 8$$

29)
$$32 \div 8$$

30)
$$40 \div 8$$

31)
$$24 \div 8$$

32)
$$40 \div 8$$

33)
$$48 \div 8$$

34)
$$16 \div 8$$

35)
$$56 \div 8$$

36)
$$80 \div 8$$

SET I Date: _____ Start: _____ Finish: _____ Score: _____

1	2	3	4	5	6
9 ÷ 9	5 4 ÷ 9	8 1 ÷ 9	7 2 ÷ 9	4 5 ÷ 9	2 7 ÷ 9

7	8	9	10	11	12
6 3 ÷ 9	9 0 ÷ 9	1 8 ÷ 9	3 6 ÷ 9	9 0 ÷ 9	1 8 ÷ 9

13	14	15	16	17	18
3 6 ÷ 9	7 2 ÷ 9	9 ÷ 9	6 3 ÷ 9	8 1 ÷ 9	4 5 ÷ 9

SET II Date: _____ Start: _____ Finish: _____ Score: _____

19	20	21	22	23	24
2 7 ÷ 9	5 4 ÷ 9	8 1 ÷ 9	7 2 ÷ 9	4 5 ÷ 9	5 4 ÷ 9

25	26	27	28	29	30
1 8 ÷ 9	2 7 ÷ 9	3 6 ÷ 9	9 ÷ 9	6 3 ÷ 9	9 0 ÷ 9

31	32	33	34	35	36
3 6 ÷ 9	6 3 ÷ 9	9 ÷ 9	9 0 ÷ 9	7 2 ÷ 9	1 8 ÷ 9

SET I   Date: _____   Start: _____   Finish: _____   Score: _____

1	2	3	4	5	6
8 1 ÷ 9	2 7 ÷ 9	4 5 ÷ 9	1 8 ÷ 9	9 0 ÷ 9	5 4 ÷ 9

7	8	9	10	11	12
3 6 ÷ 9	9 ÷ 9	6 3 ÷ 9	7 2 ÷ 9	2 7 ÷ 9	6 3 ÷ 9

13	14	15	16	17	18
3 6 ÷ 9	4 5 ÷ 9	9 0 ÷ 9	8 1 ÷ 9	5 4 ÷ 9	1 8 ÷ 9

SET II   Date: _____   Start: _____   Finish: _____   Score: _____

19	20	21	22	23	24
9 ÷ 9	7 2 ÷ 9	9 0 ÷ 9	5 4 ÷ 9	3 6 ÷ 9	4 5 ÷ 9

25	26	27	28	29	30
7 2 ÷ 9	8 1 ÷ 9	1 8 ÷ 9	6 3 ÷ 9	2 7 ÷ 9	9 ÷ 9

31	32	33	34	35	36
1 8 ÷ 9	4 5 ÷ 9	8 1 ÷ 9	9 ÷ 9	5 4 ÷ 9	9 0 ÷ 9

SET I   Date: _____   Start: _____   Finish: _____   Score: _____

1
2 7
÷ **9**

2
1 8
÷ **9**

3
9
÷ **9**

4
6 3
÷ **9**

5
8 1
÷ **9**

6
9 0
÷ **9**

7
3 6
÷ **9**

8
5 4
÷ **9**

9
4 5
÷ **9**

10
7 2
÷ **9**

11
4 5
÷ **9**

12
2 7
÷ **9**

13
9 0
÷ **9**

14
7 2
÷ **9**

15
1 8
÷ **9**

16
5 4
÷ **9**

17
6 3
÷ **9**

18
8 1
÷ **9**

SET II   Date: _____   Start: _____   Finish: _____   Score: _____

19
9
÷ **9**

20
3 6
÷ **9**

21
2 7
÷ **9**

22
9 0
÷ **9**

23
3 6
÷ **9**

24
8 1
÷ **9**

25
6 3
÷ **9**

26
1 8
÷ **9**

27
4 5
÷ **9**

28
9
÷ **9**

29
5 4
÷ **9**

30
7 2
÷ **9**

31
2 7
÷ **9**

32
4 5
÷ **9**

33
9
÷ **9**

34
8 1
÷ **9**

35
6 3
÷ **9**

36
3 6
÷ **9**

Division Facts

Practice: Dividing by 9

SET I Date: _____ Start: _____ Finish: _____ Score: _____

1	2	3	4	5	6
5 4 ÷ 9	8 1 ÷ 9	9 ÷ 9	7 2 ÷ 9	4 5 ÷ 9	9 0 ÷ 9

7	8	9	10	11	12
2 7 ÷ 9	1 8 ÷ 9	3 6 ÷ 9	6 3 ÷ 9	7 2 ÷ 9	1 8 ÷ 9

13	14	15	16	17	18
6 3 ÷ 9	2 7 ÷ 9	3 6 ÷ 9	9 ÷ 9	9 0 ÷ 9	4 5 ÷ 9

SET II Date: _____ Start: _____ Finish: _____ Score: _____

19	20	21	22	23	24
5 4 ÷ 9	8 1 ÷ 9	4 5 ÷ 9	1 8 ÷ 9	7 2 ÷ 9	6 3 ÷ 9

25	26	27	28	29	30
3 6 ÷ 9	5 4 ÷ 9	9 ÷ 9	9 0 ÷ 9	2 7 ÷ 9	8 1 ÷ 9

31	32	33	34	35	36
9 ÷ 9	2 7 ÷ 9	7 2 ÷ 9	9 0 ÷ 9	6 3 ÷ 9	3 6 ÷ 9

62 Division Facts

SET I

Date: _____ Start: _____ Finish: _____ Score: _____

1
$$2\ 4 \div 8$$

2
$$7\ 2 \div 8$$

3
$$7\ 2 \div 9$$

4
$$8\ 0 \div 8$$

5
$$1\ 6 \div 8$$

6
$$1\ 8 \div 9$$

7
$$8 \div 8$$

8
$$5\ 6 \div 8$$

9
$$9 \div 9$$

10
$$2\ 7 \div 9$$

11
$$9\ 0 \div 9$$

12
$$6\ 3 \div 9$$

13
$$3\ 6 \div 9$$

14
$$8\ 1 \div 9$$

15
$$4\ 8 \div 8$$

16
$$3\ 2 \div 8$$

17
$$4\ 0 \div 8$$

18
$$6\ 4 \div 8$$

SET II

Date: _____ Start: _____ Finish: _____ Score: _____

19
$$5\ 4 \div 9$$

20
$$4\ 5 \div 9$$

21
$$2\ 4 \div 8$$

22
$$7\ 2 \div 8$$

23
$$7\ 2 \div 9$$

24
$$8\ 0 \div 8$$

25
$$1\ 6 \div 8$$

26
$$1\ 8 \div 9$$

27
$$8 \div 8$$

28
$$5\ 6 \div 8$$

29
$$9 \div 9$$

30
$$2\ 7 \div 9$$

31
$$9\ 0 \div 9$$

32
$$6\ 3 \div 9$$

33
$$3\ 6 \div 9$$

34
$$8\ 1 \div 9$$

35
$$4\ 8 \div 8$$

36
$$3\ 2 \div 8$$

SET I   Date: _____   Start: _____   Finish: _____   Score: _____

1	2	3	4	5	6
8 ÷ 8	4 5 ÷ 9	6 3 ÷ 9	1 8 ÷ 9	4 8 ÷ 8	9 0 ÷ 9

7	8	9	10	11	12
7 2 ÷ 8	3 6 ÷ 9	1 6 ÷ 8	3 2 ÷ 8	8 0 ÷ 8	9 ÷ 9

13	14	15	16	17	18
6 4 ÷ 8	8 1 ÷ 9	4 0 ÷ 8	2 4 ÷ 8	5 6 ÷ 8	5 4 ÷ 9

SET II   Date: _____   Start: _____   Finish: _____   Score: _____

19	20	21	22	23	24
2 7 ÷ 9	7 2 ÷ 9	8 ÷ 8	4 5 ÷ 9	6 3 ÷ 9	1 8 ÷ 9

25	26	27	28	29	30
4 8 ÷ 8	9 0 ÷ 9	7 2 ÷ 8	3 6 ÷ 9	1 6 ÷ 8	3 2 ÷ 8

31	32	33	34	35	36
8 0 ÷ 8	9 ÷ 9	6 4 ÷ 8	8 1 ÷ 9	4 0 ÷ 8	2 4 ÷ 8

Division Facts

SET I   Date: _____   Start: _____   Finish: _____   Score: _____

1.
$$8\ 0$$
$$\div\quad 8$$

2.
$$5\ 4$$
$$\div\quad 9$$

3.
$$2\ 4$$
$$\div\quad 8$$

4.
$$7\ 2$$
$$\div\quad 8$$

5.
$$4\ 5$$
$$\div\quad 9$$

6.
$$3\ 2$$
$$\div\quad 8$$

7.
$$9\ 0$$
$$\div\quad 9$$

8.
$$4\ 8$$
$$\div\quad 8$$

9.
$$9$$
$$\div\quad 9$$

10.
$$5\ 6$$
$$\div\quad 8$$

11.
$$2\ 7$$
$$\div\quad 9$$

12.
$$1\ 8$$
$$\div\quad 9$$

13.
$$6\ 4$$
$$\div\quad 8$$

14.
$$8$$
$$\div\quad 8$$

15.
$$3\ 6$$
$$\div\quad 9$$

16.
$$4\ 0$$
$$\div\quad 8$$

17.
$$7\ 2$$
$$\div\quad 9$$

18.
$$1\ 6$$
$$\div\quad 8$$

SET II   Date: _____   Start: _____   Finish: _____   Score: _____

19.
$$6\ 3$$
$$\div\quad 9$$

20.
$$8\ 1$$
$$\div\quad 9$$

21.
$$8\ 0$$
$$\div\quad 8$$

22.
$$5\ 4$$
$$\div\quad 9$$

23.
$$2\ 4$$
$$\div\quad 8$$

24.
$$7\ 2$$
$$\div\quad 8$$

25.
$$4\ 5$$
$$\div\quad 9$$

26.
$$3\ 2$$
$$\div\quad 8$$

27.
$$9\ 0$$
$$\div\quad 9$$

28.
$$4\ 8$$
$$\div\quad 8$$

29.
$$9$$
$$\div\quad 9$$

30.
$$5\ 6$$
$$\div\quad 8$$

31.
$$2\ 7$$
$$\div\quad 9$$

32.
$$1\ 8$$
$$\div\quad 9$$

33.
$$6\ 4$$
$$\div\quad 8$$

34.
$$8$$
$$\div\quad 8$$

35.
$$3\ 6$$
$$\div\quad 9$$

36.
$$4\ 0$$
$$\div\quad 8$$

Division Facts

SET I

Date: _____ Start: _____ Finish: _____ Score: _____

1
$$8\ 1 \div 9$$

2
$$1\ 8 \div 9$$

3
$$6\ 3 \div 9$$

4
$$5\ 4 \div 9$$

5
$$2\ 4 \div 8$$

6
$$4\ 0 \div 8$$

7
$$2\ 7 \div 9$$

8
$$5\ 6 \div 8$$

9
$$3\ 2 \div 8$$

10
$$9\ 0 \div 9$$

11
$$9 \div 9$$

12
$$6\ 4 \div 8$$

13
$$8 \div 8$$

14
$$8\ 0 \div 8$$

15
$$7\ 2 \div 8$$

16
$$4\ 5 \div 9$$

17
$$7\ 2 \div 9$$

18
$$3\ 6 \div 9$$

SET II

Date: _____ Start: _____ Finish: _____ Score: _____

19
$$1\ 6 \div 8$$

20
$$4\ 8 \div 8$$

21
$$8\ 1 \div 9$$

22
$$1\ 8 \div 9$$

23
$$6\ 3 \div 9$$

24
$$5\ 4 \div 9$$

25
$$2\ 4 \div 8$$

26
$$4\ 0 \div 8$$

27
$$2\ 7 \div 9$$

28
$$5\ 6 \div 8$$

29
$$3\ 2 \div 8$$

30
$$9\ 0 \div 9$$

31
$$9 \div 9$$

32
$$6\ 4 \div 8$$

33
$$8 \div 8$$

34
$$8\ 0 \div 8$$

35
$$7\ 2 \div 8$$

36
$$4\ 5 \div 9$$

Division Facts

SET I　Date: _____　Start: _____　Finish: _____　Score: _____

1

$$63 \div 7$$

2

$$24 \div 6$$

3

$$36 \div 6$$

4

$$56 \div 8$$

5

$$80 \div 8$$

6

$$9 \div 9$$

7

$$27 \div 9$$

8

$$24 \div 8$$

9

$$48 \div 8$$

10

$$54 \div 9$$

11

$$14 \div 7$$

12

$$6 \div 6$$

13

$$32 \div 8$$

14

$$40 \div 8$$

15

$$81 \div 9$$

16

$$18 \div 9$$

17

$$28 \div 7$$

18

$$7 \div 7$$

SET II　Date: _____　Start: _____　Finish: _____　Score: _____

19

$$21 \div 7$$

20

$$72 \div 8$$

21

$$30 \div 6$$

22

$$60 \div 6$$

23

$$45 \div 9$$

24

$$56 \div 7$$

25

$$64 \div 8$$

26

$$48 \div 6$$

27

$$42 \div 7$$

28

$$18 \div 6$$

29

$$36 \div 9$$

30

$$72 \div 9$$

31

$$63 \div 9$$

32

$$42 \div 6$$

33

$$54 \div 6$$

34

$$70 \div 7$$

35

$$49 \div 7$$

36

$$35 \div 7$$

Division Facts

SET I   Date: _____   Start: _____   Finish: _____   Score: _____

1 2 1 ÷ 7	**2** 7 2 ÷ 8	**3** 9 ÷ 9
4 7 ÷ 7	**5** 5 6 ÷ 7	**6** 6 3 ÷ 9
7 2 7 ÷ 9	**8** 1 8 ÷ 6	**9** 5 6 ÷ 8
10 9 0 ÷ 9	**11** 4 9 ÷ 7	**12** 4 5 ÷ 9
13 4 2 ÷ 7	**14** 8 1 ÷ 9	**15** 4 0 ÷ 8
16 6 4 ÷ 8	**17** 6 ÷ 6	**18** 8 ÷ 8

SET II   Date: _____   Start: _____   Finish: _____   Score: _____

19 5 4 ÷ 6	**20** 4 8 ÷ 8	**21** 1 8 ÷ 9
22 2 4 ÷ 8	**23** 2 8 ÷ 7	**24** 2 4 ÷ 6
25 7 2 ÷ 9	**26** 3 2 ÷ 8	**27** 3 5 ÷ 7
28 3 6 ÷ 9	**29** 6 3 ÷ 7	**30** 5 4 ÷ 9
31 3 6 ÷ 6	**32** 6 0 ÷ 6	**33** 7 0 ÷ 7
34 4 2 ÷ 6	**35** 8 0 ÷ 8	**36** 1 4 ÷ 7

SET I

Date: _____ Start: _____ Finish: _____ Score: _____

1

4 2
÷ **6**

2

4 0
÷ **8**

3

7 0
÷ **7**

4

4 2
÷ **7**

5

4 5
÷ **9**

6

9
÷ **9**

7

7
÷ **7**

8

1 4
÷ **7**

9

5 4
÷ **6**

10

1 8
÷ **9**

11

6 0
÷ **6**

12

6 4
÷ **8**

13

9 0
÷ **9**

14

2 1
÷ **7**

15

1 8
÷ **6**

16

3 0
÷ **6**

17

6
÷ **6**

18

3 5
÷ **7**

SET II

Date: _____ Start: _____ Finish: _____ Score: _____

19

4 8
÷ **8**

20

3 2
÷ **8**

21

1 6
÷ **8**

22

5 6
÷ **7**

23

7 2
÷ **9**

24

5 4
÷ **9**

25

2 4
÷ **6**

26

7 2
÷ **8**

27

2 8
÷ **7**

28

4 8
÷ **6**

29

1 2
÷ **6**

30

2 7
÷ **9**

31

8
÷ **8**

32

8 0
÷ **8**

33

2 4
÷ **8**

34

8 1
÷ **9**

35

4 9
÷ **7**

36

5 6
÷ **8**

Division Facts

SET I

Date: _____ Start: _____ Finish: _____ Score: _____

1
$$48 \div 8$$

2
$$48 \div 6$$

3
$$9 \div 9$$

4
$$24 \div 8$$

5
$$64 \div 8$$

6
$$70 \div 7$$

7
$$24 \div 6$$

8
$$36 \div 9$$

9
$$45 \div 9$$

10
$$49 \div 7$$

11
$$18 \div 6$$

12
$$18 \div 9$$

13
$$54 \div 6$$

14
$$72 \div 9$$

15
$$42 \div 6$$

16
$$54 \div 9$$

17
$$56 \div 7$$

18
$$56 \div 8$$

SET II

Date: _____ Start: _____ Finish: _____ Score: _____

19
$$42 \div 7$$

20
$$7 \div 7$$

21
$$36 \div 6$$

22
$$16 \div 8$$

23
$$6 \div 6$$

24
$$30 \div 6$$

25
$$35 \div 7$$

26
$$60 \div 6$$

27
$$27 \div 9$$

28
$$28 \div 7$$

29
$$8 \div 8$$

30
$$12 \div 6$$

31
$$14 \div 7$$

32
$$63 \div 7$$

33
$$90 \div 9$$

34
$$72 \div 8$$

35
$$40 \div 8$$

36
$$80 \div 8$$

SET I Date: _____ Start: _____ Finish: _____ Score: _____

1)
$$40 \div 8$$

2)
$$18 \div 9$$

3)
$$18 \div 6$$

4)
$$54 \div 9$$

5)
$$7 \div 7$$

6)
$$28 \div 7$$

7)
$$72 \div 9$$

8)
$$42 \div 6$$

9)
$$63 \div 9$$

10)
$$30 \div 6$$

11)
$$81 \div 9$$

12)
$$21 \div 7$$

13)
$$36 \div 6$$

14)
$$14 \div 7$$

15)
$$27 \div 9$$

16)
$$60 \div 6$$

17)
$$35 \div 7$$

18)
$$56 \div 7$$

SET II Date: _____ Start: _____ Finish: _____ Score: _____

19)
$$70 \div 7$$

20)
$$48 \div 8$$

21)
$$80 \div 8$$

22)
$$45 \div 9$$

23)
$$90 \div 9$$

24)
$$48 \div 6$$

25)
$$49 \div 7$$

26)
$$24 \div 8$$

27)
$$72 \div 8$$

28)
$$6 \div 6$$

29)
$$36 \div 9$$

30)
$$8 \div 8$$

31)
$$16 \div 8$$

32)
$$64 \div 8$$

33)
$$42 \div 7$$

34)
$$24 \div 6$$

35)
$$56 \div 8$$

36)
$$63 \div 7$$

SET I Date: _____ Start: _____ Finish: _____ Score: _____

1	2	3	4	5	6
4 8 ÷ 8	2 1 ÷ 7	5 4 ÷ 9	1 8 ÷ 9	4 2 ÷ 7	4 5 ÷ 9

7	8	9	10	11	12
4 9 ÷ 7	2 7 ÷ 9	8 0 ÷ 8	8 ÷ 8	5 6 ÷ 7	1 2 ÷ 6

13	14	15	16	17	18
4 2 ÷ 6	1 6 ÷ 8	5 4 ÷ 6	8 1 ÷ 9	6 0 ÷ 6	7 2 ÷ 9

SET II Date: _____ Start: _____ Finish: _____ Score: _____

19	20	21	22	23	24
3 6 ÷ 6	7 2 ÷ 8	5 6 ÷ 8	7 ÷ 7	3 5 ÷ 7	6 4 ÷ 8

25	26	27	28	29	30
6 ÷ 6	3 6 ÷ 9	1 4 ÷ 7	6 3 ÷ 7	2 4 ÷ 6	6 3 ÷ 9

31	32	33	34	35	36
3 0 ÷ 6	4 8 ÷ 6	9 ÷ 9	4 0 ÷ 8	2 8 ÷ 7	2 4 ÷ 8

Division Facts

SET I Date: _____ Start: _____ Finish: _____ Score: _____

1	2	3	4	5	6
9 0 ÷ **1 0**	5 0 ÷ **1 0**	8 0 ÷ **1 0**	6 0 ÷ **1 0**	1 0 ÷ **1 0**	2 0 ÷ **1 0**

7	8	9	10	11	12
4 0 ÷ **1 0**	1 0 0 ÷ **1 0**	3 0 ÷ **1 0**	7 0 ÷ **1 0**	1 0 0 ÷ **1 0**	7 0 ÷ **1 0**

13	14	15	16	17	18
4 0 ÷ **1 0**	1 0 ÷ **1 0**	3 0 ÷ **1 0**	6 0 ÷ **1 0**	2 0 ÷ **1 0**	8 0 ÷ **1 0**

SET II Date: _____ Start: _____ Finish: _____ Score: _____

19	20	21	22	23	24
5 0 ÷ **1 0**	9 0 ÷ **1 0**	3 0 ÷ **1 0**	5 0 ÷ **1 0**	7 0 ÷ **1 0**	2 0 ÷ **1 0**

25	26	27	28	29	30
6 0 ÷ **1 0**	9 0 ÷ **1 0**	4 0 ÷ **1 0**	1 0 0 ÷ **1 0**	8 0 ÷ **1 0**	1 0 ÷ **1 0**

31	32	33	34	35	36
8 0 ÷ **1 0**	2 0 ÷ **1 0**	5 0 ÷ **1 0**	3 0 ÷ **1 0**	9 0 ÷ **1 0**	1 0 0 ÷ **1 0**

Division Facts

SET I   Date: _____   Start: _____   Finish: _____   Score: _____

1.
$$40 \div 10$$

2.
$$100 \div 10$$

3.
$$60 \div 10$$

4.
$$70 \div 10$$

5.
$$80 \div 10$$

6.
$$20 \div 10$$

7.
$$90 \div 10$$

8.
$$50 \div 10$$

9.
$$10 \div 10$$

10.
$$30 \div 10$$

11.
$$80 \div 10$$

12.
$$90 \div 10$$

13.
$$70 \div 10$$

14.
$$100 \div 10$$

15.
$$30 \div 10$$

16.
$$20 \div 10$$

17.
$$50 \div 10$$

18.
$$10 \div 10$$

SET II   Date: _____   Start: _____   Finish: _____   Score: _____

19.
$$60 \div 10$$

20.
$$40 \div 10$$

21.
$$60 \div 10$$

22.
$$20 \div 10$$

23.
$$40 \div 10$$

24.
$$70 \div 10$$

25.
$$100 \div 10$$

26.
$$30 \div 10$$

27.
$$10 \div 10$$

28.
$$90 \div 10$$

29.
$$80 \div 10$$

30.
$$50 \div 10$$

31.
$$90 \div 10$$

32.
$$40 \div 10$$

33.
$$60 \div 10$$

34.
$$50 \div 10$$

35.
$$30 \div 10$$

36.
$$20 \div 10$$

SET I Date: _____ Start: _____ Finish: _____ Score: _____

1
$$100 \div 10$$

2
$$60 \div 10$$

3
$$10 \div 10$$

4
$$20 \div 10$$

5
$$90 \div 10$$

6
$$80 \div 10$$

7
$$40 \div 10$$

8
$$50 \div 10$$

9
$$70 \div 10$$

10
$$30 \div 10$$

11
$$20 \div 10$$

12
$$40 \div 10$$

13
$$30 \div 10$$

14
$$90 \div 10$$

15
$$50 \div 10$$

16
$$100 \div 10$$

17
$$10 \div 10$$

18
$$70 \div 10$$

SET II Date: _____ Start: _____ Finish: _____ Score: _____

19
$$80 \div 10$$

20
$$60 \div 10$$

21
$$70 \div 10$$

22
$$50 \div 10$$

23
$$10 \div 10$$

24
$$80 \div 10$$

25
$$90 \div 10$$

26
$$40 \div 10$$

27
$$100 \div 10$$

28
$$30 \div 10$$

29
$$20 \div 10$$

30
$$60 \div 10$$

31
$$20 \div 10$$

32
$$70 \div 10$$

33
$$80 \div 10$$

34
$$10 \div 10$$

35
$$30 \div 10$$

36
$$40 \div 10$$

Division Facts

SET I   Date: _____   Start: _____   Finish: _____   Score: _____

1
$$60 \div 10$$

2
$$100 \div 10$$

3
$$80 \div 10$$

4
$$40 \div 10$$

5
$$50 \div 10$$

6
$$70 \div 10$$

7
$$30 \div 10$$

8
$$90 \div 10$$

9
$$20 \div 10$$

10
$$10 \div 10$$

11
$$100 \div 10$$

12
$$20 \div 10$$

13
$$30 \div 10$$

14
$$60 \div 10$$

15
$$70 \div 10$$

16
$$50 \div 10$$

17
$$40 \div 10$$

18
$$90 \div 10$$

SET II   Date: _____   Start: _____   Finish: _____   Score: _____

19
$$10 \div 10$$

20
$$80 \div 10$$

21
$$90 \div 10$$

22
$$30 \div 10$$

23
$$100 \div 10$$

24
$$40 \div 10$$

25
$$10 \div 10$$

26
$$70 \div 10$$

27
$$50 \div 10$$

28
$$80 \div 10$$

29
$$60 \div 10$$

30
$$20 \div 10$$

31
$$10 \div 10$$

32
$$40 \div 10$$

33
$$90 \div 10$$

34
$$50 \div 10$$

35
$$70 \div 10$$

36
$$100 \div 10$$

SET I Date: _____ Start: _____ Finish: _____ Score: _____

1
$$11 \div 11$$

2
$$99 \div 11$$

3
$$77 \div 11$$

4
$$110 \div 11$$

5
$$44 \div 11$$

6
$$88 \div 11$$

7
$$22 \div 11$$

8
$$66 \div 11$$

9
$$33 \div 11$$

10
$$55 \div 11$$

11
$$110 \div 11$$

12
$$88 \div 11$$

13
$$44 \div 11$$

14
$$77 \div 11$$

15
$$66 \div 11$$

16
$$33 \div 11$$

17
$$55 \div 11$$

18
$$22 \div 11$$

SET II Date: _____ Start: _____ Finish: _____ Score: _____

19
$$11 \div 11$$

20
$$99 \div 11$$

21
$$88 \div 11$$

22
$$99 \div 11$$

23
$$22 \div 11$$

24
$$77 \div 11$$

25
$$44 \div 11$$

26
$$55 \div 11$$

27
$$33 \div 11$$

28
$$66 \div 11$$

29
$$11 \div 11$$

30
$$110 \div 11$$

31
$$44 \div 11$$

32
$$110 \div 11$$

33
$$11 \div 11$$

34
$$99 \div 11$$

35
$$33 \div 11$$

36
$$77 \div 11$$

SET I   Date: _____   Start: _____   Finish: _____   Score: _____

1	2	3	4	5	6
1 1 0 ÷ 1 1	6 6 ÷ 1 1	3 3 ÷ 1 1	9 9 ÷ 1 1	2 2 ÷ 1 1	5 5 ÷ 1 1

7	8	9	10	11	12
8 8 ÷ 1 1	4 4 ÷ 1 1	1 1 ÷ 1 1	7 7 ÷ 1 1	6 6 ÷ 1 1	4 4 ÷ 1 1

13	14	15	16	17	18
1 1 0 ÷ 1 1	3 3 ÷ 1 1	8 8 ÷ 1 1	2 2 ÷ 1 1	9 9 ÷ 1 1	1 1 ÷ 1 1

SET II   Date: _____   Start: _____   Finish: _____   Score: _____

19	20	21	22	23	24
7 7 ÷ 1 1	5 5 ÷ 1 1	3 3 ÷ 1 1	2 2 ÷ 1 1	5 5 ÷ 1 1	8 8 ÷ 1 1

25	26	27	28	29	30
7 7 ÷ 1 1	1 1 ÷ 1 1	4 4 ÷ 1 1	1 1 0 ÷ 1 1	9 9 ÷ 1 1	6 6 ÷ 1 1

31	32	33	34	35	36
9 9 ÷ 1 1	1 1 ÷ 1 1	6 6 ÷ 1 1	5 5 ÷ 1 1	2 2 ÷ 1 1	4 4 ÷ 1 1

SET I   Date: _____   Start: _____   Finish: _____   Score: _____

1
$$99 \div 11$$

2
$$77 \div 11$$

3
$$55 \div 11$$

4
$$88 \div 11$$

5
$$11 \div 11$$

6
$$44 \div 11$$

7
$$33 \div 11$$

8
$$22 \div 11$$

9
$$110 \div 11$$

10
$$66 \div 11$$

11
$$55 \div 11$$

12
$$99 \div 11$$

13
$$22 \div 11$$

14
$$44 \div 11$$

15
$$110 \div 11$$

16
$$88 \div 11$$

17
$$33 \div 11$$

18
$$66 \div 11$$

SET II   Date: _____   Start: _____   Finish: _____   Score: _____

19
$$77 \div 11$$

20
$$11 \div 11$$

21
$$44 \div 11$$

22
$$11 \div 11$$

23
$$33 \div 11$$

24
$$55 \div 11$$

25
$$77 \div 11$$

26
$$66 \div 11$$

27
$$99 \div 11$$

28
$$88 \div 11$$

29
$$22 \div 11$$

30
$$110 \div 11$$

31
$$55 \div 11$$

32
$$11 \div 11$$

33
$$99 \div 11$$

34
$$110 \div 11$$

35
$$77 \div 11$$

36
$$44 \div 11$$

SET I   Date: _____   Start: _____   Finish: _____   Score: _____

1	2	3	4	5	6
9 9 ÷ 1 1	3 3 ÷ 1 1	4 4 ÷ 1 1	5 5 ÷ 1 1	8 8 ÷ 1 1	6 6 ÷ 1 1

7	8	9	10	11	12
1 1 0 ÷ 1 1	1 1 ÷ 1 1	7 7 ÷ 1 1	2 2 ÷ 1 1	4 4 ÷ 1 1	7 7 ÷ 1 1

13	14	15	16	17	18
5 5 ÷ 1 1	1 1 0 ÷ 1 1	3 3 ÷ 1 1	6 6 ÷ 1 1	9 9 ÷ 1 1	8 8 ÷ 1 1

SET II   Date: _____   Start: _____   Finish: _____   Score: _____

19	20	21	22	23	24
2 2 ÷ 1 1	1 1 ÷ 1 1	1 1 0 ÷ 1 1	9 9 ÷ 1 1	5 5 ÷ 1 1	8 8 ÷ 1 1

25	26	27	28	29	30
4 4 ÷ 1 1	7 7 ÷ 1 1	1 1 ÷ 1 1	6 6 ÷ 1 1	3 3 ÷ 1 1	2 2 ÷ 1 1

31	32	33	34	35	36
3 3 ÷ 1 1	8 8 ÷ 1 1	6 6 ÷ 1 1	7 7 ÷ 1 1	5 5 ÷ 1 1	1 1 ÷ 1 1

SET I Date: _____ Start: _____ Finish: _____ Score: _____

1	2	3	4	5	6
1 0 8 ÷ 1 2	8 4 ÷ 1 2	1 2 0 ÷ 1 2	2 4 ÷ 1 2	6 0 ÷ 1 2	7 2 ÷ 1 2

7	8	9	10	11	12
9 6 ÷ 1 2	4 8 ÷ 1 2	3 6 ÷ 1 2	1 2 ÷ 1 2	7 2 ÷ 1 2	1 2 0 ÷ 1 2

13	14	15	16	17	18
3 6 ÷ 1 2	8 4 ÷ 1 2	9 6 ÷ 1 2	1 2 ÷ 1 2	6 0 ÷ 1 2	4 8 ÷ 1 2

SET II Date: _____ Start: _____ Finish: _____ Score: _____

19	20	21	22	23	24
2 4 ÷ 1 2	1 0 8 ÷ 1 2	3 6 ÷ 1 2	1 2 0 ÷ 1 2	6 0 ÷ 1 2	4 8 ÷ 1 2

25	26	27	28	29	30
1 2 ÷ 1 2	2 4 ÷ 1 2	1 0 8 ÷ 1 2	8 4 ÷ 1 2	9 6 ÷ 1 2	7 2 ÷ 1 2

31	32	33	34	35	36
2 4 ÷ 1 2	1 2 ÷ 1 2	1 0 8 ÷ 1 2	6 0 ÷ 1 2	1 2 0 ÷ 1 2	4 8 ÷ 1 2

SET I Date: _____ Start: _____ Finish: _____ Score: _____

1	2	3	4	5	6
3 6 ÷ 1 2	7 2 ÷ 1 2	9 6 ÷ 1 2	1 2 0 ÷ 1 2	8 4 ÷ 1 2	1 2 ÷ 1 2

7	8	9	10	11	12
6 0 ÷ 1 2	1 0 8 ÷ 1 2	4 8 ÷ 1 2	2 4 ÷ 1 2	1 2 ÷ 1 2	2 4 ÷ 1 2

13	14	15	16	17	18
4 8 ÷ 1 2	3 6 ÷ 1 2	6 0 ÷ 1 2	9 6 ÷ 1 2	7 2 ÷ 1 2	1 0 8 ÷ 1 2

SET II Date: _____ Start: _____ Finish: _____ Score: _____

19	20	21	22	23	24
8 4 ÷ 1 2	1 2 0 ÷ 1 2	2 4 ÷ 1 2	9 6 ÷ 1 2	8 4 ÷ 1 2	1 2 ÷ 1 2

25	26	27	28	29	30
6 0 ÷ 1 2	4 8 ÷ 1 2	1 0 8 ÷ 1 2	3 6 ÷ 1 2	1 2 0 ÷ 1 2	7 2 ÷ 1 2

31	32	33	34	35	36
4 8 ÷ 1 2	6 0 ÷ 1 2	3 6 ÷ 1 2	1 2 ÷ 1 2	8 4 ÷ 1 2	7 2 ÷ 1 2

SET I   Date: _____   Start: _____   Finish: _____   Score: _____

1	2	3	4	5	6
2 4 ÷ 1 2	1 2 ÷ 1 2	1 0 8 ÷ 1 2	6 0 ÷ 1 2	1 2 0 ÷ 1 2	9 6 ÷ 1 2

7	8	9	10	11	12
4 8 ÷ 1 2	8 4 ÷ 1 2	7 2 ÷ 1 2	3 6 ÷ 1 2	6 0 ÷ 1 2	7 2 ÷ 1 2

13	14	15	16	17	18
1 0 8 ÷ 1 2	8 4 ÷ 1 2	1 2 ÷ 1 2	9 6 ÷ 1 2	2 4 ÷ 1 2	1 2 0 ÷ 1 2

SET II   Date: _____   Start: _____   Finish: _____   Score: _____

19	20	21	22	23	24
3 6 ÷ 1 2	4 8 ÷ 1 2	1 0 8 ÷ 1 2	4 8 ÷ 1 2	1 2 ÷ 1 2	2 4 ÷ 1 2

25	26	27	28	29	30
9 6 ÷ 1 2	6 0 ÷ 1 2	7 2 ÷ 1 2	8 4 ÷ 1 2	1 2 0 ÷ 1 2	3 6 ÷ 1 2

31	32	33	34	35	36
1 0 8 ÷ 1 2	4 8 ÷ 1 2	2 4 ÷ 1 2	9 6 ÷ 1 2	6 0 ÷ 1 2	1 2 ÷ 1 2

Division Facts

SET I   Date: _____   Start: _____   Finish: _____   Score: _____

(1)
48
$\div\ 12$

(2)
24
$\div\ 12$

(3)
96
$\div\ 12$

(4)
120
$\div\ 12$

(5)
36
$\div\ 12$

(6)
108
$\div\ 12$

(7)
72
$\div\ 12$

(8)
84
$\div\ 12$

(9)
60
$\div\ 12$

(10)
12
$\div\ 12$

(11)
84
$\div\ 12$

(12)
108
$\div\ 12$

(13)
36
$\div\ 12$

(14)
120
$\div\ 12$

(15)
12
$\div\ 12$

(16)
24
$\div\ 12$

(17)
96
$\div\ 12$

(18)
72
$\div\ 12$

SET II   Date: _____   Start: _____   Finish: _____   Score: _____

(19)
60
$\div\ 12$

(20)
48
$\div\ 12$

(21)
24
$\div\ 12$

(22)
120
$\div\ 12$

(23)
84
$\div\ 12$

(24)
48
$\div\ 12$

(25)
36
$\div\ 12$

(26)
60
$\div\ 12$

(27)
96
$\div\ 12$

(28)
12
$\div\ 12$

(29)
108
$\div\ 12$

(30)
72
$\div\ 12$

(31)
24
$\div\ 12$

(32)
60
$\div\ 12$

(33)
48
$\div\ 12$

(34)
12
$\div\ 12$

(35)
36
$\div\ 12$

(36)
120
$\div\ 12$

SET I Date: _____ Start: _____ Finish: _____ Score: _____

1.
$$60 \div 10$$

2.
$$70 \div 10$$

3.
$$55 \div 11$$

4.
$$40 \div 10$$

5.
$$22 \div 11$$

6.
$$110 \div 11$$

7.
$$30 \div 10$$

8.
$$120 \div 12$$

9.
$$24 \div 12$$

10.
$$99 \div 11$$

11.
$$12 \div 12$$

12.
$$90 \div 10$$

13.
$$48 \div 12$$

14.
$$77 \div 11$$

15.
$$80 \div 10$$

16.
$$100 \div 10$$

17.
$$50 \div 10$$

18.
$$20 \div 10$$

SET II Date: _____ Start: _____ Finish: _____ Score: _____

19.
$$66 \div 11$$

20.
$$36 \div 12$$

21.
$$11 \div 11$$

22.
$$88 \div 11$$

23.
$$44 \div 11$$

24.
$$96 \div 12$$

25.
$$60 \div 12$$

26.
$$84 \div 12$$

27.
$$10 \div 10$$

28.
$$72 \div 12$$

29.
$$108 \div 12$$

30.
$$33 \div 11$$

31.
$$60 \div 10$$

32.
$$70 \div 10$$

33.
$$55 \div 11$$

34.
$$40 \div 10$$

35.
$$22 \div 11$$

36.
$$110 \div 11$$

Division Facts

SET I

Date: _____ Start: _____ Finish: _____ Score: _____

1	2	3	4	5	6
4 4 ÷ 1 1	6 6 ÷ 1 1	2 4 ÷ 1 2	7 7 ÷ 1 1	1 1 0 ÷ 1 1	1 1 ÷ 1 1

7	8	9	10	11	12
8 0 ÷ 1 0	3 0 ÷ 1 0	1 0 8 ÷ 1 2	1 2 ÷ 1 2	9 6 ÷ 1 2	3 6 ÷ 1 2

13	14	15	16	17	18
5 5 ÷ 1 1	5 0 ÷ 1 0	9 0 ÷ 1 0	4 8 ÷ 1 2	3 3 ÷ 1 1	2 2 ÷ 1 1

SET II

Date: _____ Start: _____ Finish: _____ Score: _____

19	20	21	22	23	24
8 4 ÷ 1 2	8 8 ÷ 1 1	7 0 ÷ 1 0	6 0 ÷ 1 2	4 0 ÷ 1 0	2 0 ÷ 1 0

25	26	27	28	29	30
1 2 0 ÷ 1 2	7 2 ÷ 1 2	1 0 ÷ 1 0	6 0 ÷ 1 0	9 9 ÷ 1 1	1 0 0 ÷ 1 0

31	32	33	34	35	36
4 4 ÷ 1 1	6 6 ÷ 1 1	2 4 ÷ 1 2	7 7 ÷ 1 1	1 1 0 ÷ 1 1	1 1 ÷ 1 1

SET I Date: _____ Start: _____ Finish: _____ Score: _____

1
66 ÷ 11

2
88 ÷ 11

3
50 ÷ 10

4
22 ÷ 11

5
77 ÷ 11

6
100 ÷ 10

7
36 ÷ 12

8
70 ÷ 10

9
72 ÷ 12

10
40 ÷ 10

11
99 ÷ 11

12
120 ÷ 12

13
90 ÷ 10

14
12 ÷ 12

15
20 ÷ 10

16
48 ÷ 12

17
30 ÷ 10

18
84 ÷ 12

SET II Date: _____ Start: _____ Finish: _____ Score: _____

19
55 ÷ 11

20
108 ÷ 12

21
60 ÷ 12

22
44 ÷ 11

23
110 ÷ 11

24
60 ÷ 10

25
33 ÷ 11

26
24 ÷ 12

27
96 ÷ 12

28
80 ÷ 10

29
11 ÷ 11

30
10 ÷ 10

31
66 ÷ 11

32
88 ÷ 11

33
50 ÷ 10

34
22 ÷ 11

35
77 ÷ 11

36
100 ÷ 10

Division Facts

SET I Date: _____   Start: _____   Finish: _____   Score: _____

1	2	3	4	5	6
3 3 ÷ 1 1	1 0 0 ÷ 1 0	9 0 ÷ 1 0	6 6 ÷ 1 1	9 6 ÷ 1 2	7 2 ÷ 1 2

7	8	9	10	11	12
6 0 ÷ 1 0	2 4 ÷ 1 2	4 8 ÷ 1 2	4 4 ÷ 1 1	7 0 ÷ 1 0	2 0 ÷ 1 0

13	14	15	16	17	18
5 5 ÷ 1 1	1 2 ÷ 1 2	7 7 ÷ 1 1	1 1 ÷ 1 1	1 2 0 ÷ 1 2	1 1 0 ÷ 1 1

SET II Date: _____   Start: _____   Finish: _____   Score: _____

19	20	21	22	23	24
1 0 8 ÷ 1 2	6 0 ÷ 1 2	8 8 ÷ 1 1	3 0 ÷ 1 0	8 0 ÷ 1 0	1 0 ÷ 1 0

25	26	27	28	29	30
4 0 ÷ 1 0	8 4 ÷ 1 2	9 9 ÷ 1 1	5 0 ÷ 1 0	2 2 ÷ 1 1	3 6 ÷ 1 2

31	32	33	34	35	36
3 3 ÷ 1 1	1 0 0 ÷ 1 0	9 0 ÷ 1 0	6 6 ÷ 1 1	9 6 ÷ 1 2	7 2 ÷ 1 2

SET I Date: _____ Start: _____ Finish: _____ Score: _____

(1)
$$4 \div 2$$

(2)
$$21 \div 3$$

(3)
$$90 \div 9$$

(4)
$$77 \div 11$$

(5)
$$30 \div 3$$

(6)
$$64 \div 8$$

(7)
$$72 \div 8$$

(8)
$$28 \div 4$$

(9)
$$70 \div 7$$

(10)
$$50 \div 5$$

(11)
$$20 \div 10$$

(12)
$$45 \div 9$$

(13)
$$6 \div 6$$

(14)
$$16 \div 4$$

(15)
$$40 \div 8$$

(16)
$$50 \div 10$$

(17)
$$72 \div 9$$

(18)
$$54 \div 6$$

SET II Date: _____ Start: _____ Finish: _____ Score: _____

(19)
$$10 \div 5$$

(20)
$$24 \div 8$$

(21)
$$12 \div 2$$

(22)
$$18 \div 3$$

(23)
$$8 \div 4$$

(24)
$$81 \div 9$$

(25)
$$18 \div 2$$

(26)
$$8 \div 2$$

(27)
$$32 \div 8$$

(28)
$$63 \div 9$$

(29)
$$15 \div 3$$

(30)
$$25 \div 5$$

(31)
$$35 \div 7$$

(32)
$$7 \div 7$$

(33)
$$10 \div 10$$

(34)
$$63 \div 7$$

(35)
$$56 \div 8$$

(36)
$$35 \div 5$$

Division Facts

SET I Date: _____ Start: _____ Finish: _____ Score: _____

1
$$30 \div 6$$

2
$$36 \div 6$$

3
$$21 \div 7$$

4
$$72 \div 8$$

5
$$45 \div 5$$

6
$$11 \div 11$$

7
$$27 \div 3$$

8
$$7 \div 7$$

9
$$36 \div 4$$

10
$$12 \div 2$$

11
$$12 \div 6$$

12
$$12 \div 12$$

13
$$4 \div 4$$

14
$$44 \div 11$$

15
$$33 \div 11$$

16
$$15 \div 5$$

17
$$60 \div 6$$

18
$$80 \div 10$$

SET II Date: _____ Start: _____ Finish: _____ Score: _____

19
$$30 \div 3$$

20
$$90 \div 10$$

21
$$49 \div 7$$

22
$$24 \div 6$$

23
$$64 \div 8$$

24
$$60 \div 12$$

25
$$40 \div 5$$

26
$$88 \div 11$$

27
$$30 \div 10$$

28
$$5 \div 5$$

29
$$8 \div 8$$

30
$$120 \div 12$$

31
$$20 \div 5$$

32
$$99 \div 11$$

33
$$54 \div 9$$

34
$$4 \div 2$$

35
$$8 \div 4$$

36
$$72 \div 9$$

Division Facts

SET I   Date: _____   Start: _____   Finish: _____   Score: _____

1	2	3	4	5	6
2 5 ÷ **5**	7 2 ÷ **9**	9 ÷ **9**	2 2 ÷ **1 1**	6 0 ÷ **1 0**	1 2 ÷ **2**

7	8	9	10	11	12
2 4 ÷ **6**	4 9 ÷ **7**	1 0 ÷ **1 0**	4 0 ÷ **8**	1 2 ÷ **1 2**	3 ÷ **3**

13	14	15	16	17	18
1 2 ÷ **6**	5 ÷ **5**	7 ÷ **7**	6 0 ÷ **6**	4 8 ÷ **8**	2 4 ÷ **1 2**

SET II   Date: _____   Start: _____   Finish: _____   Score: _____

19	20	21	22	23	24
2 ÷ **2**	1 5 ÷ **3**	8 ÷ **8**	8 0 ÷ **1 0**	1 0 8 ÷ **1 2**	1 0 ÷ **2**

25	26	27	28	29	30
3 6 ÷ **4**	3 0 ÷ **1 0**	1 2 ÷ **3**	1 4 ÷ **7**	3 0 ÷ **3**	9 6 ÷ **1 2**

31	32	33	34	35	36
1 0 ∴ **5**	6 4 ÷ **8**	2 8 ÷ **7**	6 ÷ **2**	4 0 ÷ **4**	3 5 ÷ **5**

SET I Date: _____ Start: _____ Finish: _____ Score: _____

1	2	3	4	5	6
1 1 ÷ **1 1**	6 0 ÷ **1 2**	2 0 ÷ **2**	1 4 ÷ **2**	2 0 ÷ **5**	1 0 0 ÷ **1 0**

7	8	9	10	11	12
1 0 ÷ **2**	4 ÷ **2**	1 6 ÷ **4**	1 2 ÷ **2**	3 0 ÷ **3**	7 7 ÷ **1 1**

13	14	15	16	17	18
2 1 ÷ **7**	4 ÷ **4**	1 4 ÷ **7**	8 1 ÷ **9**	5 4 ÷ **6**	2 4 ÷ **8**

SET II Date: _____ Start: _____ Finish: _____ Score: _____

19	20	21	22	23	24
3 6 ÷ **6**	1 2 ÷ **3**	5 ÷ **5**	4 5 ÷ **5**	2 5 ÷ **5**	3 3 ÷ **1 1**

25	26	27	28	29	30
4 0 ÷ **4**	2 ÷ **2**	5 6 ÷ **7**	2 7 ÷ **3**	1 8 ÷ **3**	5 0 ÷ **1 0**

31	32	33	34	35	36
4 8 ÷ **1 2**	8 0 ÷ **8**	6 0 ÷ **1 0**	1 2 ÷ **1 2**	1 6 ÷ **2**	3 0 ÷ **6**

Division Facts

SET I   Date: _____   Start: _____   Finish: _____   Score: _____

1	2	3	4	5	6
8 0 ÷ **1 0**	4 8 ÷ **6**	9 6 ÷ **1 2**	5 4 ÷ **6**	1 6 ÷ **4**	1 1 ÷ **1 1**

7	8	9	10	11	12
6 0 ÷ **1 2**	1 2 ÷ **3**	1 0 8 ÷ **1 2**	2 ÷ **2**	2 8 ÷ **7**	1 8 ÷ **9**

13	14	15	16	17	18
3 6 ÷ **1 2**	1 0 ÷ **2**	1 0 0 ÷ **1 0**	1 2 ÷ **6**	2 0 ÷ **1 0**	4 8 ÷ **1 2**

SET II   Date: _____   Start: _____   Finish: _____   Score: _____

19	20	21	22	23	24
6 0 ÷ **6**	3 ÷ **3**	1 6 ÷ **2**	9 ÷ **3**	3 6 ÷ **9**	9 0 ÷ **9**

25	26	27	28	29	30
4 ÷ **2**	8 8 ÷ **1 1**	8 1 ÷ **9**	3 5 ÷ **5**	4 5 ÷ **5**	6 ÷ **3**

31	32	33	34	35	36
3 2 ÷ **4**	6 ÷ **2**	1 6 ÷ **8**	1 2 ÷ **1 2**	2 0 ÷ **2**	3 5 ÷ **7**

SET I Date: _____ Start: _____ Finish: _____ Score: _____

1
$$99 \div 11$$

2
$$56 \div 8$$

3
$$72 \div 9$$

4
$$90 \div 10$$

5
$$88 \div 11$$

6
$$21 \div 7$$

7
$$12 \div 6$$

8
$$18 \div 3$$

9
$$24 \div 4$$

10
$$45 \div 5$$

11
$$18 \div 9$$

12
$$30 \div 6$$

13
$$55 \div 11$$

14
$$48 \div 8$$

15
$$110 \div 11$$

16
$$54 \div 6$$

17
$$30 \div 5$$

18
$$45 \div 9$$

SET II Date: _____ Start: _____ Finish: _____ Score: _____

19
$$72 \div 12$$

20
$$108 \div 12$$

21
$$21 \div 3$$

22
$$14 \div 2$$

23
$$20 \div 2$$

24
$$4 \div 4$$

25
$$35 \div 5$$

26
$$72 \div 8$$

27
$$66 \div 11$$

28
$$80 \div 10$$

29
$$80 \div 8$$

30
$$8 \div 8$$

31
$$22 \div 11$$

32
$$14 \div 7$$

33
$$24 \div 3$$

34
$$49 \div 7$$

35
$$90 \div 9$$

36
$$50 \div 5$$

Division Facts

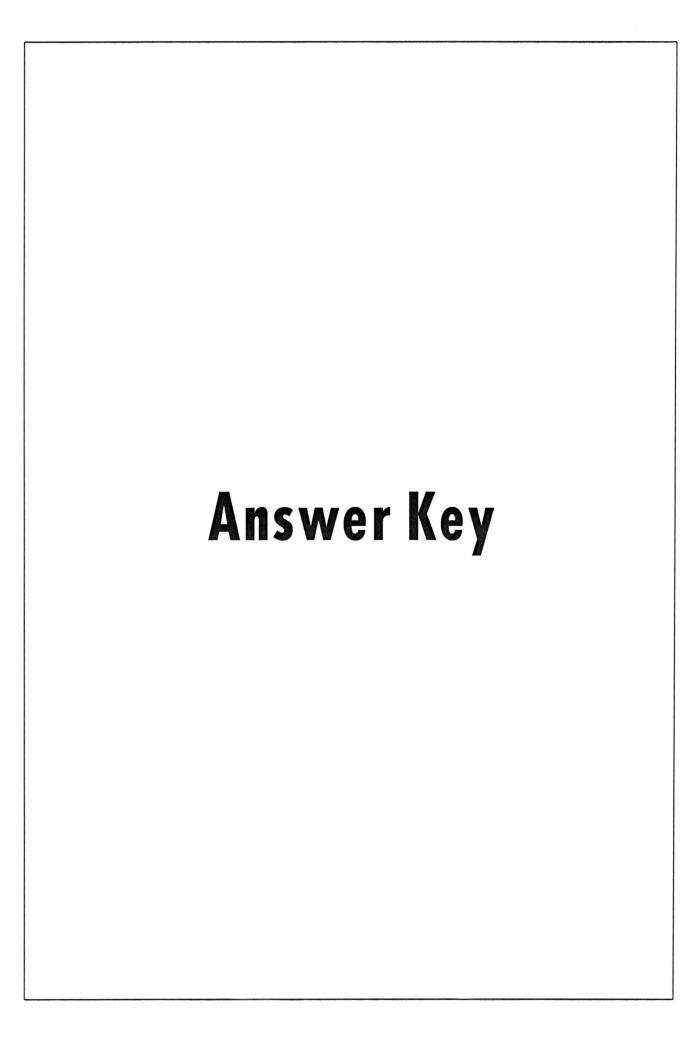

Answer Key

Page 7	34. 10	31. 3	28. 4	25. 7	22. 9	19. 6	16. 3
1. 2	35. 7	32. 5	29. 3	26. 8	23. 2	20. 7	17. 2
2. 4	36. 4	33. 9	30. 5	27. 10	24. 7	21. 1	18. 8
3. 1	**Page 8**	34. 10	31. 9	28. 9	25. 8	22. 10	19. 1
4. 10	1. 9	35. 8	32. 5	29. 2	26. 4	23. 2	20. 6
5. 6	2. 6	36. 4	33. 6	30. 6	27. 6	24. 7	21. 2
6. 9	3. 1	**Page 9**	34. 1	31. 1	28. 3	25. 9	22. 5
7. 7	4. 7	1. 10	35. 2	32. 6	29. 5	26. 8	23. 10
8. 3	5. 5	2. 2	36. 8	33. 8	30. 10	27. 5	24. 1
9. 8	6. 4	3. 6	**Page 10**	34. 10	31. 2	28. 4	25. 3
10. 5	7. 8	4. 1	1. 5	35. 7	32. 5	29. 6	26. 8
11. 1	8. 2	5. 8	2. 8	36. 9	33. 7	30. 3	27. 9
12. 4	9. 3	6. 5	3. 3	**Page 11**	34. 6	31. 10	28. 6
13. 8	10. 10	7. 7	4. 1	1. 9	35. 10	32. 3	29. 4
14. 2	11. 1	8. 4	5. 7	2. 2	36. 4	33. 1	30. 7
15. 10	12. 7	9. 9	6. 6	3. 1	**Page 12**	34. 6	31. 2
16. 7	13. 10	10. 3	7. 9	4. 7	1. 2	35. 7	32. 4
17. 9	14. 9	11. 4	8. 10	5. 4	2. 5	36. 9	33. 1
18. 6	15. 6	12. 10	9. 4	6. 8	3. 3	**Page 13**	34. 9
19. 5	16. 5	13. 1	10. 2	7. 10	4. 7	1. 10	35. 10
20. 3	17. 8	14. 2	11. 10	8. 5	5. 9	2. 4	36. 7
21. 6	18. 2	15. 5	12. 9	9. 3	6. 1	3. 2	**Page 14**
22. 10	19. 4	16. 6	13. 8	10. 6	7. 10	4. 8	1. 2
23. 2	20. 3	17. 9	14. 4	11. 2	8. 8	5. 7	2. 5
24. 5	21. 1	18. 3	15. 6	12. 5	9. 6	6. 5	3. 9
25. 3	22. 10	19. 7	16. 7	13. 4	10. 4	7. 6	4. 3
26. 9	23. 2	20. 8	17. 2	14. 1	11. 8	8. 9	5. 4
27. 7	24. 5	21. 10	18. 5	15. 6	12. 5	9. 1	6. 7
28. 4	25. 7	22. 8	19. 1	16. 10	13. 1	10. 3	7. 10
29. 1	26. 6	23. 6	20. 3	17. 9	14. 9	11. 4	8. 1
30. 8	27. 3	24. 7	21. 4	18. 3	15. 3	12. 9	9. 6
31. 5	28. 4	25. 1	22. 3	19. 8	16. 10	13. 5	10. 8
32. 8	29. 8	26. 2	23. 5	20. 7	17. 4	14. 7	11. 9
33. 2	30. 9	27. 9	24. 1	21. 1	18. 2	15. 10	12. 2

Division Facts

13. 4
14. 5
15. 7
16. 8
17. 6
18. 10
19. 3
20. 1
21. 8
22. 7
23. 3
24. 5
25. 4
26. 1
27. 2
28. 10
29. 6
30. 9
31. 2
32. 1
33. 10
34. 7
35. 6
36. 9

Page 15
1. 5
2. 1
3. 10
4. 2
5. 3
6. 6
7. 4
8. 8
9. 9

10. 7
11. 6
12. 9
13. 7
14. 5
15. 1
16. 8
17. 2
18. 4
19. 3
20. 10
21. 1
22. 5
23. 3
24. 4
25. 9
26. 6
27. 8
28. 10
29. 2
30. 7
31. 8
32. 2
33. 9
34. 10
35. 3
36. 6

Page 16
1. 7
2. 2
3. 1
4. 5
5. 3
6. 10

7. 6
8. 9
9. 4
10. 8
11. 3
12. 2
13. 7
14. 1
15. 6
16. 8
17. 4
18. 10
19. 9
20. 5
21. 7
22. 3
23. 1
24. 2
25. 4
26. 9
27. 5
28. 8
29. 10
30. 6
31. 2
32. 5
33. 6
34. 3
35. 9
36. 8

Page 17
1. 2
2. 7
3. 2

4. 5
5. 9
6. 3
7. 7
8. 1
9. 8
10. 3
11. 10
12. 6
13. 6
14. 1
15. 4
16. 8
17. 4
18. 5
19. 10
20. 9
21. 2
22. 7
23. 2
24. 5
25. 9
26. 3
27. 7
28. 1
29. 8
30. 3
31. 10
32. 6
33. 6
34. 1
35. 4
36. 8

Page 18
1. 4
2. 7
3. 6
4. 10
5. 4
6. 3
7. 8
8. 8
9. 10
10. 5
11. 9
12. 9
13. 5
14. 2
15. 7
16. 1
17. 6
18. 2
19. 1
20. 3
21. 4
22. 7
23. 6
24. 10
25. 4
26. 3
27. 8
28. 8
29. 10
30. 5
31. 9
32. 9
33. 5

34. 2
35. 7
36. 1

Page 19
1. 8
2. 10
3. 3
4. 5
5. 1
6. 2
7. 2
8. 1
9. 5
10. 9
11. 10
12. 6
13. 4
14. 8
15. 7
16. 3
17. 7
18. 9
19. 4
20. 6
21. 8
22. 10
23. 3
24. 5
25. 1
26. 2
27. 2
28. 1
29. 5
30. 9

31. 10
32. 6
33. 4
34. 8
35. 7
36. 3

Page 20
1. 6
2. 1
3. 8
4. 6
5. 3
6. 7
7. 9
8. 8
9. 2
10. 10
11. 4
12. 5
13. 7
14. 4
15. 9
16. 10
17. 5
18. 1
19. 3
20. 2
21. 6
22. 1
23. 8
24. 6
25. 3
26. 7
27. 9

28. 8
29. 2
30. 10
31. 4
32. 5
33. 7
34. 4
35. 9
36. 10

Page 21
1. 8
2. 4
3. 7
4. 5
5. 10
6. 10
7. 7
8. 5
9. 6
10. 1
11. 6
12. 2
13. 8
14. 4
15. 9
16. 3
17. 2
18. 3
19. 1
20. 9
21. 8
22. 4
23. 7
24. 5

Division Facts

25. 10	22. 1	19. 5	16. 1	13. 6	10. 6	7. 1	4. 2
26. 10	23. 8	20. 9	17. 8	14. 3	11. 8	8. 6	5. 7
27. 7	24. 4	21. 4	18. 6	15. 10	12. 3	9. 7	6. 5
28. 5	25. 6	22. 8	19. 7	16. 1	13. 10	10. 10	7. 1
29. 6	26. 7	23. 6	20. 2	17. 9	14. 2	11. 8	8. 8
30. 1	27. 1	24. 7	21. 8	18. 8	15. 9	12. 4	9. 3
31. 6	28. 7	25. 2	22. 1	19. 4	16. 6	13. 2	10. 10
32. 2	29. 3	26. 1	23. 4	20. 7	17. 1	14. 9	11. 2
33. 8	30. 6	27. 9	24. 7	21. 8	18. 7	15. 3	12. 8
34. 4	31. 2	28. 3	25. 10	22. 4	19. 5	16. 7	13. 9
35. 9	32. 2	29. 5	26. 3	23. 5	20. 4	17. 6	14. 4
36. 3	33. 4	30. 10	27. 5	24. 2	21. 9	18. 5	15. 3
Page 22	34. 5	31. 5	28. 6	25. 9	22. 1	19. 10	16. 5
1. 8	35. 5	32. 2	29. 2	26. 1	23. 4	20. 1	17. 6
2. 1	36. 9	33. 6	30. 9	27. 3	24. 6	21. 2	18. 10
3. 8	**Page 23**	34. 7	31. 4	28. 6	25. 8	22. 7	19. 7
4. 4	1. 8	35. 1	32. 10	29. 10	26. 10	23. 1	20. 1
5. 6	2. 2	36. 3	33. 3	30. 7	27. 7	24. 4	21. 7
6. 7	3. 7	**Page 24**	34. 5	31. 9	28. 3	25. 6	22. 1
7. 1	4. 10	1. 8	35. 1	32. 2	29. 5	26. 10	23. 5
8. 7	5. 5	2. 4	36. 8	33. 10	30. 2	27. 3	24. 3
9. 3	6. 9	3. 6	**Page 25**	34. 1	31. 1	28. 5	25. 6
10. 6	7. 4	4. 9	1. 9	35. 4	32. 5	29. 9	26. 8
11. 2	8. 1	5. 5	2. 6	36. 6	33. 9	30. 8	27. 10
12. 2	9. 3	6. 3	3. 1	**Page 26**	34. 7	31. 6	28. 9
13. 4	10. 6	7. 10	4. 7	1. 8	35. 3	32. 2	29. 2
14. 5	11. 7	8. 2	5. 4	2. 1	36. 10	33. 1	30. 4
15. 5	12. 6	9. 7	6. 3	3. 9	**Page 27**	34. 3	31. 5
16. 9	13. 8	10. 1	7. 10	4. 7	1. 3	35. 4	32. 4
17. 10	14. 4	11. 9	8. 8	5. 2	2. 4	36. 8	33. 3
18. 3	15. 10	12. 3	9. 2	6. 5	3. 5	**Page 28**	34. 7
19. 10	16. 3	13. 4	10. 5	7. 3	4. 2	1. 4	35. 9
20. 9	17. 2	14. 10	11. 2	8. 10	5. 8	2. 9	36. 8
21. 8	18. 1	15. 5	12. 5	9. 4	6. 9	3. 6	

Division Facts

Page 29	34. 9	31. 7	28. 3	25. 6	22. 4	19. 6	16. 8
1. 2	35. 3	32. 1	29. 2	26. 7	23. 7	20. 7	17. 10
2. 3	36. 4	33. 10	30. 6	27. 5	24. 2	21. 5	18. 7
3. 6	**Page 30**	34. 9	31. 1	28. 2	25. 7	22. 3	19. 3
4. 8	1. 9	35. 4	32. 7	29. 6	26. 5	23. 3	20. 10
5. 7	2. 7	36. 8	33. 9	30. 3	27. 10	24. 5	21. 7
6. 10	3. 5	**Page 31**	34. 1	31. 10	28. 8	25. 4	22. 1
7. 9	4. 2	1. 3	35. 4	32. 1	29. 5	26. 9	23. 10
8. 5	5. 10	2. 7	36. 10	33. 4	30. 1	27. 7	24. 4
9. 1	6. 4	3. 6	**Page 32**	34. 2	31. 3	28. 8	25. 3
10. 4	7. 8	4. 5	1. 5	35. 10	32. 9	29. 9	26. 2
11. 10	8. 3	5. 8	2. 7	36. 8	33. 6	30. 2	27. 1
12. 7	9. 6	6. 10	3. 1	**Page 33**	34. 4	31. 1	28. 3
13. 5	10. 1	7. 2	4. 9	1. 6	35. 1	32. 8	29. 6
14. 3	11. 8	8. 3	5. 6	2. 4	36. 9	33. 10	30. 6
15. 4	12. 3	9. 2	6. 7	3. 7	**Page 34**	34. 1	31. 1
16. 8	13. 2	10. 6	7. 5	4. 2	1. 5	35. 10	32. 2
17. 6	14. 10	11. 1	8. 2	5. 7	2. 3	36. 2	33. 3
18. 2	15. 9	12. 7	9. 6	6. 5	3. 3	**Page 35**	34. 7
19. 1	16. 7	13. 9	10. 3	7. 10	4. 5	1. 8	35. 2
20. 9	17. 6	14. 1	11. 10	8. 8	5. 4	2. 10	36. 6
21. 7	18. 4	15. 4	12. 1	9. 5	6. 9	3. 9	**Page 36**
22. 8	19. 5	16. 10	13. 4	10. 1	7. 7	4. 9	1. 2
23. 1	20. 1	17. 5	14. 2	11. 3	8. 8	5. 4	2. 2
24. 2	21. 6	18. 8	15. 10	12. 9	9. 9	6. 9	3. 9
25. 3	22. 10	19. 4	16. 8	13. 6	10. 2	7. 2	4. 8
26. 10	23. 3	20. 9	17. 4	14. 4	11. 1	8. 7	5. 4
27. 9	24. 2	21. 3	18. 8	15. 1	12. 8	9. 5	6. 4
28. 6	25. 8	22. 7	19. 9	16. 9	13. 10	10. 4	7. 8
29. 4	26. 7	23. 6	20. 3	17. 2	14. 1	11. 4	8. 5
30. 5	27. 9	24. 5	21. 5	18. 8	15. 10	12. 8	9. 10
31. 10	28. 5	25. 8	22. 7	19. 3	16. 2	13. 8	10. 1
32. 1	29. 1	26. 10	23. 1	20. 10	17. 4	14. 5	11. 2
33. 8	30. 4	27. 2	24. 9	21. 6	18. 6	15. 6	12. 1

13. 9	10. 4	7. 7	4. 4	**Page 40**	34. 8	31. 3	28. 2
14. 5	11. 10	8. 5	5. 6	1. 10	35. 3	32. 4	29. 6
15. 6	12. 9	9. 8	6. 9	2. 9	36. 6	33. 7	30. 7
16. 6	13. 6	10. 2	7. 4	3. 5	**Page 41**	34. 2	31. 6
17. 8	14. 7	11. 4	8. 1	4. 9	1. 1	35. 6	32. 1
18. 10	15. 10	12. 4	9. 8	5. 8	2. 6	36. 9	33. 9
19. 4	16. 4	13. 7	10. 1	6. 4	3. 2	**Page 42**	34. 5
20. 3	17. 3	14. 1	11. 2	7. 3	4. 5	1. 4	35. 7
21. 5	18. 8	15. 9	12. 5	8. 10	5. 10	2. 7	36. 10
22. 5	19. 8	16. 3	13. 7	9. 4	6. 7	3. 2	**Page 43**
23. 7	20. 6	17. 5	14. 7	10. 1	7. 3	4. 10	1. 4
24. 7	21. 10	18. 3	15. 6	11. 1	8. 4	5. 5	2. 1
25. 7	22. 4	19. 9	16. 3	12. 9	9. 9	6. 8	3. 6
26. 6	23. 7	20. 5	17. 8	13. 7	10. 8	7. 6	4. 8
27. 3	24. 2	21. 8	18. 8	14. 3	11. 6	8. 1	5. 10
28. 6	25. 9	22. 5	19. 10	15. 5	12. 9	9. 9	6. 7
29. 10	26. 8	23. 1	20. 5	16. 2	13. 8	10. 3	7. 5
30. 3	27. 1	24. 3	21. 6	17. 8	14. 5	11. 4	8. 9
31. 4	28. 6	25. 3	22. 6	18. 6	15. 3	12. 2	9. 3
32. 9	29. 1	26. 1	23. 10	19. 10	16. 7	13. 7	10. 2
33. 1	30. 9	27. 2	24. 2	20. 4	17. 10	14. 6	11. 8
34. 3	31. 7	28. 2	25. 4	21. 4	18. 1	15. 8	12. 6
35. 8	32. 9	29. 2	26. 4	22. 1	19. 4	16. 9	13. 4
36. 9	33. 5	30. 8	27. 9	23. 9	20. 2	17. 3	14. 10
Page 37	34. 6	31. 9	28. 2	24. 7	21. 10	18. 10	15. 7
1. 8	35. 5	32. 8	29. 5	25. 5	22. 2	19. 5	16. 2
2. 2	36. 10	33. 7	30. 7	26. 7	23. 1	20. 1	17. 3
3. 3	**Page 38**	34. 7	31. 3	27. 7	24. 9	21. 4	18. 1
4. 2	1. 6	35. 4	32. 3	28. 2	25. 7	22. 3	19. 5
5. 3	2. 6	36. 10	33. 3	29. 10	26. 4	23. 10	20. 9
6. 2	3. 9	**Page 39**	34. 5	30. 2	27. 8	24. 1	21. 2
7. 3	4. 6	1. 10	35. 9	31. 8	28. 5	25. 8	22. 7
8. 4	5. 10	2. 8	36. 7	32. 3	29. 3	26. 9	23. 4
9. 1	6. 10	3. 9		33. 6	30. 6	27. 5	24. 5

Division Facts

25. 9	22. 4	19. 6	16. 5	13. 1	10. 4	7. 4	4. 1
26. 6	23. 1	20. 2	17. 10	14. 3	11. 2	8. 10	5. 4
27. 8	24. 8	21. 5	18. 7	15. 9	12. 7	9. 2	6. 2
28. 1	25. 10	22. 7	19. 3	16. 6	13. 10	10. 8	7. 8
29. 3	26. 5	23. 4	20. 6	17. 2	14. 4	11. 6	8. 3
30. 10	27. 2	24. 8	21. 10	18. 7	15. 1	12. 3	9. 9
31. 6	28. 3	25. 2	22. 1	19. 5	16. 3	13. 5	10. 3
32. 5	29. 6	26. 1	23. 3	20. 8	17. 5	14. 5	11. 2
33. 4	30. 9	27. 10	24. 2	21. 1	18. 8	15. 1	12. 6
34. 1	31. 2	28. 3	25. 7	22. 5	19. 9	16. 9	13. 6
35. 9	32. 5	29. 6	26. 5	23. 8	20. 6	17. 7	14. 7
36. 8	33. 1	30. 9	27. 9	24. 4	21. 8	18. 6	15. 5
Page 44	34. 7	31. 6	28. 8	25. 7	22. 1	19. 8	16. 4
1. 1	35. 8	32. 1	29. 4	26. 3	23. 6	20. 3	17. 9
2. 2	36. 3	33. 10	30. 6	27. 6	24. 4	21. 4	18. 5
3. 5	**Page 45**	34. 8	31. 5	28. 10	25. 3	22. 1	19. 10
4. 4	1. 6	35. 7	32. 2	29. 2	26. 2	23. 10	20. 10
5. 3	2. 1	36. 2	33. 9	30. 9	27. 5	24. 2	21. 1
6. 8	3. 10	**Page 46**	34. 3	31. 10	28. 7	25. 9	22. 7
7. 10	4. 4	1. 10	35. 8	32. 4	29. 10	26. 7	23. 8
8. 9	5. 3	2. 9	36. 10	33. 7	30. 9	27. 4	24. 1
9. 6	6. 2	3. 2	**Page 47**	34. 2	31. 4	28. 10	25. 4
10. 7	7. 9	4. 5	1. 3	35. 1	32. 8	29. 2	26. 2
11. 6	8. 8	5. 8	2. 6	36. 5	33. 6	30. 8	27. 8
12. 7	9. 5	6. 6	3. 5	**Page 48**	34. 9	31. 6	28. 3
13. 5	10. 7	7. 4	4. 2	1. 1	35. 5	32. 3	29. 9
14. 4	11. 9	8. 7	5. 7	2. 7	36. 7	33. 5	30. 3
15. 2	12. 3	9. 3	6. 9	3. 8	**Page 49**	34. 5	31. 2
16. 1	13. 4	10. 1	7. 4	4. 9	1. 4	35. 1	32. 6
17. 10	14. 8	11. 1	8. 10	5. 6	2. 1	36. 9	33. 6
18. 8	15. 1	12. 8	9. 8	6. 2	3. 10	**Page 50**	34. 7
19. 9	16. 10	13. 2	10. 1	7. 5	4. 2	1. 1	35. 5
20. 3	17. 7	14. 4	11. 10	8. 3	5. 9	2. 7	36. 4
21. 7	18. 5	15. 9	12. 4	9. 10	6. 7	3. 8	

Page 51

1. 2
2. 5
3. 7
4. 6
5. 6
6. 4
7. 1
8. 9
9. 5
10. 8
11. 10
12. 4
13. 3
14. 2
15. 10
16. 8
17. 1
18. 3
19. 9
20. 7
21. 2
22. 5
23. 7
24. 6
25. 6
26. 4
27. 1
28. 9
29. 5
30. 8
31. 10
32. 4
33. 3
34. 2
35. 10
36. 8

Page 52

1. 2
2. 9
3. 9
4. 5
5. 7
6. 3
7. 10
8. 3
9. 4
10. 4
11. 6
12. 2
13. 8
14. 10
15. 5
16. 1
17. 8
18. 1
19. 6
20. 7
21. 2
22. 9
23. 9
24. 5
25. 7
26. 3
27. 10
28. 3
29. 4
30. 4
31. 6
32. 2
33. 8
34. 10
35. 5
36. 1

Page 53

1. 9
2. 5
3. 6
4. 2
5. 1
6. 8
7. 1
8. 10
9. 7
10. 8
11. 9
12. 7
13. 10
14. 4
15. 2
16. 5
17. 6
18. 3
19. 3
20. 4
21. 9
22. 5
23. 6
24. 2
25. 1
26. 8
27. 1
28. 10
29. 7
30. 8
31. 9
32. 7
33. 10
34. 4
35. 2
36. 5

Page 54

1. 6
2. 5
3. 10
4. 7
5. 3
6. 4
7. 1
8. 1
9. 9
10. 8
11. 3
12. 4
13. 6
14. 9
15. 5
16. 10
17. 2
18. 8
19. 2
20. 7
21. 6
22. 5
23. 10
24. 7
25. 3
26. 4
27. 1
28. 1
29. 9
30. 8
31. 3
32. 4
33. 6
34. 9
35. 5
36. 10

Page 55

1. 6
2. 2
3. 4
4. 10
5. 8
6. 7
7. 5
8. 9
9. 3
10. 1
11. 2
12. 8
13. 9
14. 6
15. 1
16. 7
17. 4
18. 5
19. 3
20. 10
21. 9
22. 4
23. 7
24. 1
25. 3
26. 2
27. 10
28. 6
29. 5
30. 8
31. 10
32. 1
33. 2
34. 6
35. 9
36. 5

Page 56

1. 3
2. 7
3. 9
4. 10
5. 6
6. 1
7. 5
8. 8
9. 2
10. 4
11. 7
12. 6
13. 8
14. 5
15. 10
16. 2
17. 3
18. 1
19. 4
20. 9
21. 7
22. 2
23. 4
24. 8
25. 5
26. 3
27. 10
28. 6
29. 1
30. 9
31. 10
32. 3
33. 6
34. 4
35. 1
36. 2

Page 57

1. 10
2. 4
3. 8
4. 5
5. 7
6. 2
7. 3
8. 9
9. 6
10. 1
11. 9
12. 6
13. 5
14. 1
15. 3
16. 10
17. 4
18. 7
19. 2
20. 8
21. 6
22. 2
23. 3
24. 8
25. 4
26. 5
27. 1
28. 9
29. 10
30. 7
31. 3
32. 10
33. 1
34. 8
35. 2
36. 6

Page 58

1. 4
2. 10
3. 3
4. 2
5. 7
6. 5
7. 9
8. 8
9. 6
10. 1
11. 3
12. 5

Division Facts

13. 9	10. 4	7. 4	4. 7	**Page 62**	34. 10	31. 10	28. 4
14. 10	11. 10	8. 1	5. 9	1. 6	35. 7	32. 7	29. 2
15. 4	12. 2	9. 7	6. 10	2. 9	36. 4	33. 4	30. 4
16. 6	13. 4	10. 8	7. 4	3. 1	**Page 63**	34. 9	31. 10
17. 8	14. 8	11. 3	8. 6	4. 8	1. 3	35. 6	32. 1
18. 1	15. 1	12. 7	9. 5	5. 5	2. 9	36. 4	33. 8
19. 2	16. 7	13. 4	10. 8	6. 10	3. 8	**Page 64**	34. 9
20. 7	17. 9	14. 5	11. 5	7. 3	4. 10	1. 1	35. 5
21. 1	18. 5	15. 10	12. 3	8. 2	5. 2	2. 5	36. 3
22. 2	19. 3	16. 9	13. 10	9. 4	6. 2	3. 7	**Page 65**
23. 7	20. 6	17. 6	14. 8	10. 7	7. 1	4. 2	1. 10
24. 10	21. 9	18. 2	15. 2	11. 8	8. 7	5. 6	2. 6
25. 6	22. 8	19. 1	16. 6	12. 2	9. 1	6. 10	3. 3
26. 3	23. 5	20. 8	17. 7	13. 7	10. 3	7. 9	4. 9
27. 8	24. 6	21. 10	18. 9	14. 3	11. 10	8. 4	5. 5
28. 9	25. 2	22. 6	19. 1	15. 4	12. 7	9. 2	6. 4
29. 4	26. 3	23. 4	20. 4	16. 1	13. 4	10. 4	7. 10
30. 5	27. 4	24. 5	21. 3	17. 10	14. 9	11. 10	8. 6
31. 3	28. 1	25. 8	22. 10	18. 5	15. 6	12. 1	9. 1
32. 5	29. 7	26. 9	23. 4	19. 6	16. 4	13. 8	10. 7
33. 6	30. 10	27. 2	24. 9	20. 9	17. 5	14. 9	11. 3
34. 2	31. 4	28. 7	25. 7	21. 5	18. 8	15. 5	12. 2
35. 7	32. 7	29. 3	26. 2	22. 2	19. 6	16. 3	13. 8
36. 10	33. 1	30. 1	27. 5	23. 8	20. 5	17. 7	14. 1
Page 59	34. 10	31. 2	28. 1	24. 7	21. 3	18. 6	15. 4
1. 1	35. 8	32. 5	29. 6	25. 4	22. 9	19. 3	16. 5
2. 6	36. 2	33. 9	30. 8	26. 6	23. 8	20. 8	17. 8
3. 9	**Page 60**	34. 1	31. 3	27. 1	24. 10	21. 1	18. 2
4. 8	1. 9	35. 6	32. 5	28. 10	25. 2	22. 5	19. 7
5. 5	2. 3	36. 10	33. 1	29. 3	26. 2	23. 7	20. 9
6. 3	3. 5	**Page 61**	34. 9	30. 9	27. 1	24. 2	21. 10
7. 7	4. 2	1. 3	35. 7	31. 1	28. 7	25. 6	22. 6
8. 10	5. 10	2. 2	36. 4	32. 3	29. 1	26. 10	23. 3
9. 2	6. 6	3. 1		33. 8	30. 3	27. 9	24. 9

25. 5	22. 2	19. 3	16. 8	13. 10	10. 7	7. 8	4. 2
26. 4	23. 7	20. 9	17. 1	14. 3	11. 3	8. 7	5. 6
27. 10	24. 6	21. 5	18. 1	15. 3	12. 2	9. 7	6. 5
28. 6	25. 3	22. 10	19. 9	16. 5	13. 9	10. 5	7. 7
29. 1	26. 5	23. 5	20. 6	17. 1	14. 8	11. 9	8. 3
30. 7	27. 3	24. 8	21. 2	18. 5	15. 7	12. 3	9. 10
31. 3	28. 7	25. 8	22. 3	19. 6	16. 6	13. 6	10. 1
32. 2	29. 4	26. 8	23. 4	20. 4	17. 8	14. 2	11. 8
33. 8	30. 10	27. 6	24. 4	21. 2	18. 7	15. 3	12. 2
34. 1	31. 1	28. 3	25. 8	22. 8	19. 6	16. 10	13. 7
35. 4	32. 8	29. 4	26. 4	23. 8	20. 1	17. 5	14. 2
36. 5	33. 1	30. 8	27. 5	24. 6	21. 6	18. 8	15. 9
Page 66	34. 10	31. 7	28. 4	25. 4	22. 2	19. 10	16. 9
1. 9	35. 9	32. 7	29. 9	26. 9	23. 1	20. 6	17. 10
2. 2	36. 5	33. 9	30. 6	27. 4	24. 5	21. 10	18. 8
3. 7	**Page 67**	34. 10	31. 6	28. 8	25. 5	22. 5	19. 6
4. 6	1. 9	35. 7	32. 10	29. 2	26. 10	23. 10	20. 9
5. 3	2. 4	36. 5	33. 10	30. 3	27. 3	24. 8	21. 7
6. 5	3. 6	**Page 68**	34. 7	31. 1	28. 4	25. 7	22. 1
7. 3	4. 7	1. 3	35. 10	32. 10	29. 1	26. 3	23. 5
8. 7	5. 10	2. 9	36. 2	33. 3	30. 2	27. 9	24. 8
9. 4	6. 1	3. 1	**Page 69**	34. 9	31. 2	28. 1	25. 1
10. 10	7. 3	4. 1	1. 7	35. 7	32. 9	29. 4	26. 4
11. 1	8. 3	5. 8	2. 5	36. 7	33. 10	30. 1	27. 2
12. 8	9. 6	6. 7	3. 10	**Page 70**	34. 9	31. 2	28. 9
13. 1	10. 6	7. 3	4. 6	1. 6	35. 5	32. 8	29. 4
14. 10	11. 2	8. 3	5. 5	2. 8	36. 10	33. 6	30. 7
15. 9	12. 1	9. 7	6. 1	3. 1	**Page 71**	34. 4	31. 5
16. 5	13. 4	10. 10	7. 1	4. 3	1. 5	35. 7	32. 8
17. 8	14. 5	11. 7	8. 2	5. 8	2. 2	36. 9	33. 1
18. 4	15. 9	12. 5	9. 9	6. 10	3. 3	**Page 72**	34. 5
19. 2	16. 2	13. 6	10. 2	7. 4	4. 6	1. 6	35. 4
20. 6	17. 4	14. 9	11. 10	8. 4	5. 1	2. 3	36. 3
21. 9	18. 1	15. 5	12. 8	9. 5	6. 4	3. 6	

Page 73

1. 9
2. 5
3. 8
4. 6
5. 1
6. 2
7. 4
8. 10
9. 3
10. 7
11. 10
12. 7
13. 4
14. 1
15. 3
16. 6
17. 2
18. 8
19. 5
20. 9
21. 3
22. 5
23. 7
24. 2
25. 6
26. 9
27. 4
28. 10
29. 8
30. 1
31. 8
32. 2
33. 5
34. 3
35. 9
36. 10

Page 74

1. 4
2. 10
3. 6
4. 7
5. 8
6. 2
7. 9
8. 5
9. 1
10. 3
11. 8
12. 9
13. 7
14. 10
15. 3
16. 2
17. 5
18. 1
19. 6
20. 4
21. 6
22. 2
23. 4
24. 7
25. 10
26. 3
27. 1
28. 9
29. 8
30. 5
31. 9
32. 4
33. 6
34. 5
35. 3
36. 2

Page 75

1. 10
2. 6
3. 1
4. 2
5. 9
6. 8
7. 4
8. 5
9. 7
10. 3
11. 2
12. 4
13. 3
14. 9
15. 5
16. 10
17. 1
18. 7
19. 8
20. 6
21. 7
22. 5
23. 1
24. 8
25. 9
26. 4
27. 10
28. 3
29. 2
30. 6
31. 2
32. 7
33. 8
34. 1
35. 3
36. 4

Page 76

1. 6
2. 10
3. 8
4. 4
5. 5
6. 7
7. 3
8. 9
9. 2
10. 1
11. 10
12. 2
13. 3
14. 6
15. 7
16. 5
17. 4
18. 9
19. 1
20. 8
21. 9
22. 3
23. 10
24. 4
25. 1
26. 7
27. 5
28. 8
29. 6
30. 2
31. 1
32. 4
33. 9
34. 5
35. 7
36. 10

Page 77

1. 1
2. 9
3. 7
4. 10
5. 4
6. 8
7. 2
8. 6
9. 3
10. 5
11. 10
12. 8
13. 4
14. 7
15. 6
16. 3
17. 5
18. 2
19. 1
20. 9
21. 8
22. 9
23. 2
24. 7
25. 4
26. 5
27. 3
28. 6
29. 1
30. 10
31. 4
32. 10
33. 1
34. 9
35. 3
36. 7

Page 78

1. 10
2. 6
3. 3
4. 9
5. 2
6. 5
7. 8
8. 4
9. 1
10. 7
11. 6
12. 4
13. 10
14. 3
15. 8
16. 2
17. 9
18. 1
19. 7
20. 5
21. 3
22. 2
23. 5
24. 8
25. 7
26. 1
27. 4
28. 10
29. 9
30. 6
31. 9
32. 1
33. 6
34. 5
35. 2
36. 4

Page 79

1. 9
2. 7
3. 5
4. 8
5. 1
6. 4
7. 3
8. 2
9. 10
10. 6
11. 5
12. 9
13. 2
14. 4
15. 10
16. 8
17. 3
18. 6
19. 7
20. 1
21. 4
22. 1
23. 3
24. 5
25. 7
26. 6
27. 9
28. 8
29. 2
30. 10
31. 5
32. 1
33. 9
34. 10
35. 7
36. 4

Page 80

1. 9
2. 3
3. 4
4. 5
5. 8
6. 6
7. 10
8. 1
9. 7
10. 2
11. 4
12. 7

13. 5
14. 10
15. 3
16. 6
17. 9
18. 8
19. 2
20. 1
21. 10
22. 9
23. 5
24. 8
25. 4
26. 7
27. 1
28. 6
29. 3
30. 2
31. 3
32. 8
33. 6
34. 7
35. 5
36. 1

Page 81

1. 9
2. 7
3. 10
4. 2
5. 5
6. 6
7. 8
8. 4
9. 3
10. 1
11. 6
12. 10
13. 3
14. 7
15. 8
16. 1
17. 5
18. 4
19. 2
20. 9
21. 3
22. 10
23. 5
24. 4
25. 1
26. 2
27. 9
28. 7
29. 8
30. 6
31. 2
32. 1
33. 9
34. 5
35. 10
36. 4

Page 82

1. 3
2. 6
3. 8
4. 10
5. 7
6. 1
7. 5
8. 9
9. 4
10. 2
11. 1
12. 2
13. 4
14. 3
15. 5
16. 8
17. 6
18. 9
19. 7
20. 10
21. 2
22. 8
23. 7
24. 1
25. 5
26. 4
27. 9
28. 3
29. 10
30. 6
31. 4
32. 5
33. 3
34. 1
35. 7
36. 6

Page 83

1. 2
2. 1
3. 9
4. 5
5. 10
6. 8
7. 4
8. 7
9. 6
10. 3
11. 5
12. 6
13. 9
14. 7
15. 1
16. 8
17. 2
18. 10
19. 3
20. 4
21. 9
22. 4
23. 1
24. 2
25. 8
26. 5
27. 6
28. 7
29. 10
30. 3
31. 9
32. 4
33. 2
34. 8
35. 5
36. 1

Page 84

1. 4
2. 2
3. 8
4. 10
5. 3
6. 9
7. 6
8. 7
9. 5
10. 1
11. 7
12. 9
13. 3
14. 10
15. 1
16. 2
17. 8
18. 6
19. 5
20. 4
21. 2
22. 10
23. 7
24. 4
25. 3
26. 5
27. 8
28. 1
29. 9
30. 6
31. 2
32. 5
33. 4
34. 1
35. 3
36. 10

Page 85

1. 6
2. 7
3. 5
4. 4
5. 2
6. 10
7. 3
8. 10
9. 2
10. 9
11. 1
12. 9
13. 4
14. 7
15. 8
16. 10
17. 5
18. 2
19. 6
20. 3
21. 1
22. 8
23. 4
24. 8
25. 5
26. 7
27. 1
28. 6
29. 9
30. 3
31. 6
32. 7
33. 5
34. 4
35. 2
36. 10

Page 86

1. 4
2. 6
3. 2
4. 7
5. 10
6. 1
7. 8
8. 3
9. 9
10. 1
11. 8
12. 3
13. 5
14. 5
15. 9
16. 4
17. 3
18. 2
19. 7
20. 8
21. 7
22. 5
23. 4
24. 2
25. 10
26. 6
27. 1
28. 6
29. 9
30. 10
31. 4
32. 6
33. 2
34. 7
35. 10
36. 1

Page 87

1. 6
2. 8
3. 5
4. 2
5. 7
6. 10
7. 3
8. 7
9. 6
10. 4
11. 9
12. 10
13. 9
14. 1
15. 2
16. 4
17. 3
18. 7
19. 5
20. 9
21. 5
22. 4
23. 10
24. 6

Division Facts

25. 3	22. 3	19. 2	16. 3	13. 2	10. 6	7. 5	4. 9
26. 2	23. 8	20. 3	17. 10	14. 1	11. 10	8. 4	5. 8
27. 8	24. 1	21. 6	18. 8	15. 1	12. 7	9. 9	6. 3
28. 8	25. 4	22. 6	19. 10	16. 10	13. 3	10. 1	7. 2
29. 1	26. 7	23. 2	20. 9	17. 6	14. 1	11. 4	8. 6
30. 1	27. 9	24. 9	21. 7	18. 2	15. 2	12. 2	9. 6
31. 6	28. 5	25. 9	22. 4	19. 1	16. 9	13. 3	10. 9
32. 8	29. 2	26. 4	23. 8	20. 5	17. 9	14. 5	11. 2
33. 5	30. 3	27. 4	24. 5	21. 1	18. 3	15. 10	12. 5
34. 2	31. 3	28. 7	25. 8	22. 8	19. 6	16. 2	13. 5
35. 7	32. 10	29. 5	26. 8	23. 9	20. 4	17. 2	14. 6
36. 10	33. 9	30. 5	27. 3	24. 5	21. 1	18. 4	15. 10
Page 88	34. 6	31. 5	28. 1	25. 9	22. 9	19. 10	16. 9
1. 3	35. 8	32. 1	29. 1	26. 3	23. 5	20. 1	17. 6
2. 10	36. 6	33. 1	30. 10	27. 4	24. 3	21. 8	18. 5
3. 9	**Page 89**	34. 9	31. 4	28. 2	25. 10	22. 3	19. 6
4. 6	1. 2	35. 7	32. 9	29. 10	26. 1	23. 4	20. 9
5. 8	2. 7	36. 7	33. 6	30. 8	27. 8	24. 10	21. 7
6. 6	3. 10	**Page 90**	34. 2	31. 2	28. 9	25. 2	22. 7
7. 6	4. 7	1. 5	35. 2	32. 8	29. 6	26. 8	23. 10
8. 2	5. 10	2. 6	36. 8	33. 4	30. 5	27. 9	24. 1
9. 4	6. 8	3. 3	**Page 91**	34. 3	31. 4	28. 7	25. 7
10. 4	7. 9	4. 9	1. 5	35. 10	32. 10	29. 9	26. 9
11. 7	8. 7	5. 9	2. 8	36. 7	33. 6	30. 2	27. 6
12. 2	9. 10	6. 1	3. 1	**Page 92**	34. 1	31. 8	28. 8
13. 5	10. 10	7. 9	4. 2	1. 1	35. 8	32. 3	29. 10
14. 1	11. 2	8. 1	5. 6	2. 5	36. 5	33. 2	30. 1
15. 7	12. 5	9. 9	6. 6	3. 10	**Page 93**	34. 1	31. 2
16. 1	13. 1	10. 6	7. 4	4. 7	1. 8	35. 10	32. 2
17. 10	14. 4	11. 2	8. 7	5. 4	2. 8	36. 5	33. 8
18. 10	15. 5	12. 1	9. 1	6. 10	3. 8	**Page 94**	34. 7
19. 9	16. 5	13. 1	10. 5	7. 5	4. 9	1. 9	35. 10
20. 5	17. 8	14. 4	11. 1	8. 2	5. 4	2. 7	36. 10
21. 8	18. 9	15. 3	12. 1	9. 4	6. 1	3. 8	

This page is intentionally left blank

This page is intentionally left blank

CPSIA information can be obtained
at www.ICGtesting.com
Printed in the USA
LVOW09s1232081117
555055LV00007B/10/P

9 781536 971477